# THE LIVING CLOCK

OTHER BOOKS ON BIOLOGICAL RHYTHMS
BY JOHN PALMER

*Biological Clock: Two views* Academic Press, New York, (with Frank A. Brown, Jr. and J. Woodland Hastings)

*Biological Clocks in Marine Organisms: the Control of Physiological and Behavioral Tidal Rhythms* Wiley (InterScience), New York.

*An Introduction to Biological Rhythms* Academic Press, New York.

*The Biological Rhythms and Clocks of Intertidal Animals* Oxford University Press. New York.

# THE LIVING CLOCK

*The Orchestrator of Biological Rhythms*

*John D. Palmer*

OXFORD

UNIVERSITY PRESS

2002

# OXFORD
UNIVERSITY PRESS

Oxford   New York
Auckland   Bangkok   Buenos Aires   Cape Town
Chennai   Dar es Salaam   Delhi   Hong Kong   Istanbul   Karachi   Kolkata
Kuala Lumpur   Madrid   Melbourne   Mexico City   Mumbai   Nairobi
Sao Paolo   Shanghai   Singapore   Taipei   Tokyo   Toronto

and associated companies in
Berlin   Ibadan

Copyright © 2002 by Oxford University Press, Inc.

Published by Oxford University Press, Inc.
198 Madison Avenue, New York, New York 10016

Oxford is a registered trademark of Oxford University Press

Library of Congress Cataloging-in-Publication Data
Palmer, John D., 1932–
    The living clock / John D. Palmer.
       p.   cm.
    Includes bibliographical references and index.
    ISBN 0-19-514340-X
    1. Chronobiology—Popular works.   I. Title.
QP84.6 P35   2002
571.7'7—dc21

                                        2001034011


        1   3   5   7   9   8   6   4   2
Printed in the United States of America
on acid free paper

*To my wife, Carla*

# ACKNOWLEDGMENTS

To the many who have advised and helped me in so many ways in this endeavor, I thank all of you. Particular recognition and thanks go to Harold Dowse, Lawrence Feldman, Judith Goodenough, Marc Roussel, Theodore Sargent, the entire library staff of the Marine Biological Laboratory at Woods Hole Massachusetts, the anonymous reviewers, and to my editor, Kirk Jensen. Blame them for any mistakes (just kidding).

Special thanks to the Digital Corporation (R.I.P.) for having the forethought to give my ancient computer (used to write this book) the wherewithal to sail unhampered into the new millennium.

# Contents

# PREFACE

I sat down at a window seat in the last row of the plane and grumbled at the certainty that I would be served dinner last. I opened my attache case, pushed aside a giant land crab there in a plastic bag, and took out some work to do. Under my seat was a box containing 150 crabs that I was also taking back to my university for use in research. Soon a woman in a business suit—a CEO, I learned later—sat down in the aisle seat and said, "What a day, a drink will really hit the spot." I had been in the field for weeks and agreed in silence. The last passenger to board was a coed radiating heat from trying to squeeze a month of beach time into a week-long spring break in Florida. She took the seat between the business woman and myself.

The CEO could not resist asking about the scratching sounds coming from the box under my seat. I explained that crabs have within their bodies a living clock, in essence, roughly the same one humans possess, and that I was a biologist studying their behavior. By the time the drink cart finally reached us I was deep in discussion with the people in earshot around me.

I explained to the group that the crab in my attache case was special (at least to me). In the lab, in the absence of any clues to the passage of time, it sat around almost motionlessly all day, but at 6 P.M. every evening it began to run around in its container and continued to do so much of the night. By using its living clock it was able to precisely meet this starting time. In recognition of his impressive punctuality I had named him "Big Ben."

We landed at Kennedy Airport at 7 P.M., and forgetting about the crab, I opened my attache case to replace the work I had taken out earlier. At 6 P.M. Big Ben had, of course, torn out of the plastic bag. With his two ominous claws, each half the size of a lobster's, open wide and at the ready, he charged toward the coed, and amazingly, she did a *grand jeté* over people into the aisle. Big Ben followed, scrambling over the CEO's lap with me in pursuit. A colorful ending to my lecture, one dramatically driving home the point of living-clock-controlled behavior.

The story above is—in a roundabout way—a reason for the existence of this book. As a marine biologist I travel to lots of places around the world, and fellow travellers usually ask what I do for a living. When I tell them, their first question is, "Do you really get paid to do that?" The answer is yes, but then they want more and more information. They are fascinated to hear that each of them have living clocks within their bodies, as do all plants and animals. People's uniform interest in the subject, plus the importance of the topic, finally convinced me that I should try to share this information with the reading public—thus the volume you now hold in your hands.

The book is written for nonscientists. While having had one good introductory course in biology will ease your understanding, it is not a requirement. The presentation style is light, and biological terminology is held to an absolute minimum. The chapters are short and can pretty much stand alone; thus readers can stop and contemplate as desired. The book is somewhat autobiographical: I've worked in this field for nearly 40 years, and I still maintain my fascination and enthusiasm for the subject.

These clocks within us are very important. They determine the basic rates (for example, intensity or vigor) at which a great many of our fundamental body processes run. They do this automatically so we are oblivious to their presence until we fly to new time zones; then jet lag symptoms signal our timepieces' existence, actions, and temporary "unhappiness."

All plants and animals have these timepieces; living clock control is ubiquitous throughout the organic world. Thus, if we are to truly understand living things we must comprehend their clocks. Until this task is accomplished, humanity will lack an exceedingly important key to understanding life. This book is an attempt to give readers a leg up in that direction.

The clock works so subtly that its existence was not foreseen. It was finally recognized in the early part of this century, but only sporadic interest followed for many years. In fact, the majority of scientists became convinced fully of the clock's existence not very long ago. Biologists still do not understand completely how the timepiece works, but we are rapidly closing in on it; and when the escapement mechanism is finally deciphered, a Nobel Prize should be the reward for the discoverer(s). Lest you think I exaggerate, here is a second opinion. Once each year the American Association for the Advancement of Science publishes the "Breakthroughs of the Year." For two consecutive years new discoveries about the living clock made that organization's short, most impressive list.

Clearly, scientists have come to see the importance of the living clock. Hopefully the public will follow their lead—they should: This clock is synonymous with life itself.

*John D. Palmer*
*University of Massachusetts*
*Amherst, Massachusetts*

# INTRODUCTION TO RHYTHMS AND CLOCKS

## THE RHYTHMIC MIGRATIONS OF EUGLENA
### A MICROSCOPIC PLANT/ANIMAL

I stood on the famous cast-iron Clifton suspension bridge—built in 1845 by Isambard Kingdom Brunel—looking down on the River Avon. A newcomer to this part of southern England would never suspect that the river below was the terminus of the same quaint, willow-tree margined stream that flows though Shakespeare's Stratford. Here, where it empties into the Bristol Channel, it is (or hopefully was, the year was 1962) a highly polluted, turbid effluent from upstream humanity. Another difference is that this stretch of the river experiences the second highest tidal exchange on earth: something like 43 feet depending on the day of the month. As I stared down from my lofty overlook, standing in the exact spot where modern bungee jumping would be born in 1979—with the mini-car traffic barely missing my backsides—I watched the tide recede. Its exit first exposed a miasmatic black mud on each side of the ever-narrowing channel of water. Next the stench of the ooze wafted up to me. Eventually, this channel, that a few hours earlier was deep enough to allow large ocean-going vessels to pass along to the city of Bristol just upstream, was now only a few feet wide. Additionally spectacular, the terrible black river muds mysteriously transmogrified before my eyes into a lush verdant green! As the tide began to rise again, just before it would inundate a new level of mud, the green color vanished. Was this section of the river inhabited by playful leprechauns that had surfed in from the Irish Sea?

The next day I stood by river's edge under the bridge waiting for the tide to recede. Again, as it did before, the surface of the exposed mud slowly turned green. The smell was terrible, but I donned rubber hip boots and with more than a little trepidation, stepped into the newly exposed mephitic ooze. Walking was very difficult because with each step I sunk down to my knee caps, and then the suction made it almost impossible to pull my foot out without leaving the boot stuck there. I worried that I might easily lose my balance and fall face down in the disgusting river bottom.[1] I pressed ahead and collected some of the green-covered mud in a Petri dish and retreated back up to my car. There I found a group of locals who had come out of the Red Lion Pub to watch my derring-do; as if I were the paid entertainment, some of them seemed disappointed at not seeing a pratfall. But they were friendly pub regulars, and once I got my disgustingly mud-frosted boots off, and they learned that I was a "Yank," they invited me into the pub for a pint of bitter.

Thus refreshed, I returned to my lab and examined the mud under a microscope. This was a peek-and-shriek viewing: Everything in the muck was not dead and rotting as the smell certainly suggested; in fact, the opposite was true, it was teeming with life. The main inhabitant was a plant/animal called *Euglena,* an organism that consists of a single cell that can undergo photosynthesis just like other plants, also has a mouth (called a gullet) through which it ingests living things smaller than itself; it even has an eye called a stigma.

It can also locomote by worming its way through the slimy mud. All these abilities are packed into a 130 by 25 micron dollop of protoplasm (a micron is one thousandth of a millimeter).

I soon learned the daily routine of this humble organism's life. When the tide is in, the *Euglena* bury themselves in the mud where they cannot be washed away. But when the tide recedes they ascend up and out of the mud to sit on the surface for some photosynthetic sunbathing. They are so abundant on the surface they are forced to pile on top of one another and form a layer thick enough to produce a visible verdant cover that can be seen from the bridge, 248 feet overhead. As the tide slowly returns to fill the channel, just before it inundates a new level of the shoreline, the *Euglena* reburrow into their nether world, and therefore do not wash away.

Because this behavior pattern is repeated with such beat-like regularity it is described (at least by the more romantic and classically trained scientists) as a rhythm. In spite of the relatively short distance travelled, the pattern is considered a migration. However, it is not like those of birds and caribou over and across the landscape, but up and down. Thus, this *Euglena* is said to display a vertical-migration rhythm.

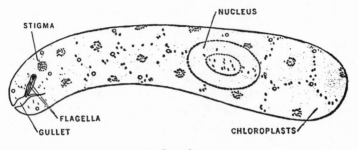

*Euglena* obtusa

My new finding became financially profitable for me. After memorizing the schedule of tides I would carefully collect *Euglena*-bearing mud samples in Petri dishes from locations successively farther down the river bank as the tide receded. I would then place these dishes, with the *Euglena* sitting on the mud surface, on the sidewalk outside of the Red Lion and invite interested patrons to join me in a game. I'd say to the sports among them, "All these dishes will turn black at different times. You pick any one of them you wish and I'll tell you within plus or minus five minutes when the dish's green color will disappear. If I am right, each of you who bet will owe me a pint of bitter; if I am wrong, I'll buy you a pint." It was so obvious that such a trick could not be done that most of the patrons, and even the publican, were more than willing, at first, to take advantage of the foolish yank so deranged that he seemed to enjoy wallowing in Avon mud. However, the outcome was that I essentially drank for nothing the entire year I spent on the project—I could always find someone new who would gamble on such a sure thing. Additionally, I estimate that many pints are still owed me, because English pints are 20 ounces, and my liver could not have survived that year had I fully collected my bets. When I would return home after a day's "work" my wife often mentioned that I seemed so happy, "You must really enjoy living in England."

Since I cannot bet with you readers, I'll tell you how the ruse was accomplished. But first, some background. The tide, when it returns, rises up the shore slowly. Just before it reaches a group of *Euglena*, they bury themselves. This progressive disappearance up the slope could be clearly observed when looking down from the Brunel Bridge. In my first attempt at understanding this, I hypothesized that underground water from the returning tide must actually reach the *Euglena* before the obvious front of surface water did. For example, perhaps the water was racing ahead of that front by capillarity through the interstices between the mud grains. The experiment designed to test this was an easy one: Isolate a sample of mud in a glass Petri dish so this subterranean advance water could never get to them. Smugly (because I was sure I knew the answer in advance) I almost did not try the experiment, but when I did, to my surprise I found that the cells in the dish still submerged in synchrony with their immediate neighbors left at that level of the shoreline!

Now, returning to my gambling shenanigans, each of the *Euglena* and mud samples on the sidewalk had been carefully collected from different levels of the shoreline, and all I had to do was remember which was which and what time each dish would have been covered by the incoming tide— an easy cerebration unless one had consumed too much beer. Every so often I would play hustler and intentionally give the wrong answer, which by then usually meant only that a winner could cancel one of the pints he already owed me.

Unless you are one of my old companions from the Red Lion days and are now concentrating only on how to get even, you must be wondering how the *Euglena* "knew" what time to reburrow. How can something as simple in construction as a single cell, devoid of a complex brain and unable to own and read a Rolex like myself (make that a Timex), perform such a feat? As hard as it may be to believe, in addition to being a plant that can both locomote and eat and an animal that can photosynthesize, *Euglena* has a living clock within its body that it uses to direct its temporal life! How can I be sure of such a claim?

"It's elementary," as Holmes would say. I just took *Euglena*-bearing mud samples into the laboratory and placed them in a room where the temperature was unchanging, the overhead light was left on all the time, and, of course, there were no tides. Thus, they were now in a truly unnatural place where there were no environmental clues as to the time of day or the state of the tide. Then, at one-hour intervals, I would observe whether the *Euglena* were on the mud surface or not. So as to be more specific than just recording green or non-green, I developed a method—affectionately called the toilet-paper technique—of rather precisely quantifying the changes. Tiny squares of identical sized bathroom tissue were placed on the mud surface, and when the *Euglena* ascended out of the mud in search of light, they foolishly moved right into the interstices of the paper. Then, periodically, a single square of paper was removed, and the *Euglena* entrapped therein washed out and were then counted under the microscope. The only brilliant part of the design was choosing English toilet paper (1962 vintage), which really had no other practical application because it was as coarse as sandpaper, waterproof, and tough enough to be reusable. I found that on the first day of isolation in the lab, the *Euglena* underwent a rhythm just as did their old neighbors back on the shoreline. Clearly, each one had a tiny clock built right into its protoplasm!

*The Water-Jet Travels of the Commuter Diatom*

Needless to say, I learned a great deal more about Euglena behavior that year (I didn't spend all my time drinking beer), but rather than carry on, I'll move on a year or two to myself in Cape Cod, Massachusetts, standing on the firm sandy bottom of Barnstable Harbor, barefooted and in a bathing suit. It is 6 A.M., the sun has just risen and the tide is beginning to recede. The breeze is at just the right temperature for my attire, and the gulls are circling overhead making those choking sounds as if they must have just eaten a clam with its shell still on. I think to myself in this clamorous solitude, "It's incredible that I get paid to be here doing this."

As the water retreated from around my feet, a sandy bottom strewn with broken shells was exposed. Unsurprisingly, at first the sand was sandcolored, but then splotches of golden brown begin appearing here and there on its surface. After my experience in England, I expected to find that living things were responsible for these color changes, and that was the case here too: Under the microscope I found another kind of motile plant, this time a diatom.

Diatoms are single-cell organisms, uniquely curious in that they enclose their golden-brown protoplasm inside a transparent glassy container. There are some pores and channels penetrating the showcase wall through which mucus may be extruded. This propels the alga jetboat through the sand grains style. Looking down on these minute diatoms through the microscope—they are only a fraction of the length of a *Euglena*—and seeing them scoot over and around the enormous (to them) sand grains is only slightly different than peering down at the street traffic from the top of New York's Empire State Building: lots of hustle and bustle below by dot-sized cars that augment the viewer's vertigo. Like *Euglena*, the diatoms emerged onto the surface sediments during daytime low tides, but remained interred at other times.

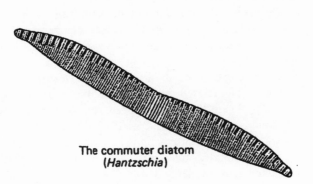

**The commuter diatom
(*Hantzschia*)**

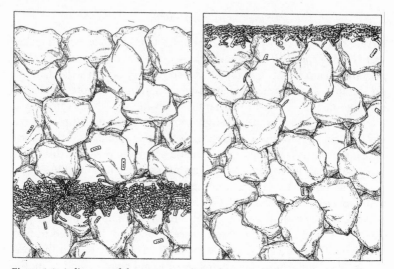

*Figure 1.1. A diagram of the commuter diatom's travels during daytime low tides. At the left the alga are seen as a layer of cells less than a millimeter below the surface of the sand; it is here where they reside during high tides and at night. The view at the right shows them piled up on the surface. The rough boulders are actually sand grains.*

Also like *Euglena*, they piled on top of one another on the surface making their presence apparent to the naked eye. Sometimes they can even be heard! An old naturalist observed that the diatoms were so abundant on the surface that their photosynthetic activity was distinctly audible as a gentle sizzling . . . and that the sand was frothy with bubbles of gas, presumably by oxygen given off by them. (Happily for humans and other non-plants, one of the by-products of photosynthesis is the life-sustaining oxygen that plants expel into the environment.)

Fortunately, I still had some English toilet paper left, and I used it to quantify the diatom's vertical migration rhythm. After transfer to the lab, we found that the commuter diatom's rhythm was a bona fide clock-controlled one.

To study the rhythm under controlled conditions, we carried sand containing diatom samples 28 miles to the Marine Biological Laboratory in Woods Hole, Mass. (an institution considered, at least by we members, to be the most famous marine lab in the world), and placed them in an incubator in which the temperature and light could be held exactly constant. Needless to say, the tides were also precluded. Nevertheless, although detained in an ambience devoid of all time cues, day after day the tiny alga's peregrination rhythm persisted with such accuracy that some people working at the marine lab became pests: As word spread about this experiment, when others in the lab wanted to go to Barnstable Harbor to collect clams (to eat, not to experiment upon) they found it easier to peek into the incubator and see the progress of the commuter diatom's rhythm than to use the published tide tables, which always required some corrective calculations. (On average, tides repeat themselves at 12 hour and 25 minute intervals, meaning that an afternoon tide on one day will occur 50 minutes later the next afternoon.)

I will digress for a paragraph to shed some doubt on what would otherwise be an obvious hypothesis. Considering the energy expended by such a minute organism on travels over such long migrations (speaking relatively), makes one feel that Mother Nature must have intended the effort to fulfil some need. Since the alga requires sunlight to live, it seemed obvious that the commuter diatom should emerge at low tide to undergo maximum photosynthesis so as to synthesize sufficient carbohydrates to sustain it through at least the nighttime and high tide submergences. While that scenario appeared so unambiguous that it didn't require verification, we ran the test anyway. This was fortunate because it destroyed our hypothesis . . . another case of speculation slain by facts. Please pardon a moment's return to high school chemistry: The carbohydrates synthesized during photosynthesis are made from carbon dioxide ($CO_2$) and water ($H_2O$). An easy way to measure the rate of photosynthesis is to offer the diatom $CO_2$ in which the C has been replaced by a radioactive version (the $^{14}C$ isotope); how "hot" the tiny plants become is a measure of their photosynthetic rate. We made these measurements and to our great surprise found that the commuter diatom is what is called a shade species, just like the shorter trees and shrubs in a rain forest that need relatively less light for maximum photosynthesis than the stately giants towering overhead and shading them below. Our study discovered that at Barnstable Harbor all the light necessary for the diatom's requirement penetrated two millimeters into the sand (well below the depth to which they burrow), and that the cells, by fully exposing themselves on the surface sands, caused their photosynthetic rate to be *inhibited* by 14 percent. We still do not have a good explanation as to why the wee commuters undergo these daily ups and downs.

*Sex in a Single French Worm:*
*The Rhythmic Reproduction of/convoluta*

I will give one last example of a tide-associated, vertical-migration rhythm. On the beach directly in front of the Roscoff Marine Laboratory in the Brittany part of western France lives a little flattened worm named *Convoluta roscoffensis* (sorry, there is no common name).

The worms are only four millimeters long, but billions live wound together within the sand at night and during high tides, or on the surface of the sand during daytime low tides. That gregariousness is reflected in the animal's name that roughly translates from the Latin to a tangled mass living in Roscoff. The worm is green because it has green algal cells actually living and reproducing within its body tissues. When the worm is at the sunbathing position of its migration rhythm the alga can photosynthesize to replenish its food stores. Then, with no concern for fair play or morality, the worm cavalierly digests some of its internal tenants as a source of food. One might think of this lifestyle as a kind of an internal, herbivorous self-cannibalism.

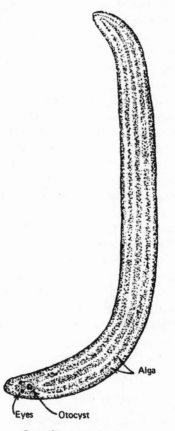

Convoluta roscofensis

The alga need not suffer this fatal outcome, for it can, and does, live quite well sans worm; but as far as the worm's survival is concerned the algal symbiont's presence is an obligate component of the worm. If *Convoluta* is placed in total darkness for several days its alga die, causing the animal to starve to death.

The toilet-paper technique cannot be used to study this worm's travels for two reasons: *Convoluta* is too large to squeeze into the interstices of the tenacious tissue, and as an investigator approaches the animals on the beach they quickly burrow. While it has eyes they are not of the image-forming type, so the worm does not submerge because it sees you coming. It ducks thanks to a warning signaled by its organ of equilibrium, called an otocyst, that senses the vibrations emanating from one's falling footsteps. But *Convoluta* is still easy to collect because it gives off a terrible stench (like rotting fish). One need only collect some barren smelly sand and can be sure that it is still wormy. The sand is then returned to the lab in petri dishes, and as long as the lab personnel tiptoe around, the worms undergo their vertical migration rhythm in synchrony with their feral cousins back on the beach. This is thus a manifestation of the presence of a clock in the multicellular animal.

The verdant color of the worms contrasts nicely with the white beach sands, and with a bit of practice, an investigator can rather accurately estimate the size of a worm population by the greenness of the sand. When such estimates are done daily, it is noticed that the size of the population shrinks and swells at two-week intervals. This fortnightly rhythm in population density is not created in the way one might surmise. The change in population size has to do with reproduction, and the animals do this in a way thought curious to humans. Procreation in this animal is by *ne plus ultra* sex. Each worm has both male and female reproductive systems, so that when any two worms feel the urge to get physical they exchange sperm by mutual, simultaneous insemination, and each starts on the road to becoming pregnant mother-fathers. In the summer this ritual repeats every other week and terminates just after their posterior ends have become swollen with embryos. Birthing consists of each worm simply casting off its entire *derrière* and leaving it buried in the sand where the bum disintegrates, liberating the developing young who must then fend for themselves. The adult population continues its commutations to the surface in undiminished numbers, but at first the group appears only half as numerous as before because now it consists of only worm front ends.

## RECAP

It is probably worthwhile to ruminate a bit on the content of this first chapter. My intent here was to introduce the concept of organismic rhythms and their creation and governance by the actions of living clocks, the latter being a kind of timepiece, as you will see, that exists in all living things—ourselves included. The basic periods of these clock-controlled rhythms are tidal (12.4 hours to complete a cycle), daily (24 hours), fortnightly (about 15 days), and even annual (365 days). In this chapter I deliberately chose as actors so-called simpler organisms (a misnomer if there ever was one) like single-cell algae and a lowly flatworm, because they not only display interesting rhythms, but also have other fascinating features like living in a glass house (the commuter diatom) or being chimeric plant/animals (*Euglena* and *Convoluta*). I also have pointed out that hypotheses are often wrong; in other words, nature isn't always what it seems to be (there is a frequently used guideline called Ockham's Razor that states that the simplest explanation is the best one; but we should stop to consider that in all likelihood Friar William of Ockham did not shave). Lastly, puzzling out the escapement mechanism of the living clock is a fundamental, very important problem not only facing biologists, but all of us: Much of the daily lives of humans is directed by our clocks, making the subject deeply personal to each of us. The next three chapters will focus on human rhythms.

All of the findings described in this short chapter could fall into the category of—depending on the attitude of the reader—interesting anecdotes that may amuse and/or fascinate; or, might well raise the question, "So what?" The experimental results described above and elsewhere in the book are examples of pure science, the opposite of applied science. The latter comprise investigations specifically directed toward producing some social good, such as the prevention of polio; while pure science is an attempt to understand the basic workings of nature. Some humanitarians may be upset at the amounts of money spent by governments, universities, and corporations on pure research, but that approach often ends up being the most useful for humanity. For example, both Drs. Jonas Salk and Albert Sabin developed vaccines that could, if all governments would support their use, rid the entire planet of polio; but it was Dr. John Enders whose pure science provided the biological basics used by the two notables in their quest for a vaccine, and it was Enders, not them, who received the Nobel Prize!

Thus far, chronobiologists (as we in the field refer to ourselves) have learned that all living things, including humans, contain living clocks that inexorably guide the important temporal aspects of their lives. We are now working hard to discover the intricacies of the clockworks. There is no question (at least in my mind) that the individuals who eventually decode this mechanism will become candidates for a Nobel Prize. As you will soon see, this basic knowledge will even save human lives.

# HUMAN RHYTHMS: BASIC PROCESSES

## HUMANS AS AN EXPERIMENTAL SUBJECT

In a word: difficult. The main reason in two words: hates isolation. Dungeons, "the box," "the hole," solitary confinement, "Go to your room," et cetra, have been, since we first became civilized, a standard form of punishment. Social animals like to be around their own kind.

And there is the rub: In the study of biological rhythms, because of the important influence of the day/night cycle on rhythms, all subjects—even the human ones—must be isolated in constant conditions for extended periods of time. There aren't many normal people who, even for a price and a free place to eat and sleep for six weeks, are willing to endure the loneliness. Thus, using those who are willing to volunteer often has its limitations because a good experiment should involve only subjects who are representative of the population as a whole. On top of that, as anyone who has ever kept a lover knows, they are expensive and difficult to maintain: They can be fastidious about what they eat and their creature comforts. Even when willing volunteers are found, if they begin to miss a loved one, become fidgety over whether the stock market is going up or down, or whatever, their psychological state often disrupts their physiological rhythms and spoils the experiment.

I will digress momentarily to demonstrate how unexpected problems arise. A sculptor, who for the sake of maintaining anonymity I will call Mick Angelo, volunteered to live alone for several weeks in an experimental isolation chamber while wired up to recording instruments that would monitor his many activities. Mick even chose the most difficult kind of situation to endure: toughing it out in complete darkness. That overtly curious choice was based on an experiment of his own creation that he planned to do while in isolation. In return for serving as a human guinea pig, he negotiated to have an enormous mound of wet clay stacked in the darkness with him so he could try sculpting by feel only, never being able to actually see what he was creating (sounds interesting, doesn't it). In this kind of experiment the subject has to prepare in advance: Mick had to practice feeling his way around his new quarters, preparing hot meals in the pitch darkness, bathing safely, being sure not to leave the toilet seat up, and all the rest. Once the experiment begins, a subject is not allowed to talk to his overseers outside the chamber, so need-requests are passed as notes slipped under the door; fortunately Mr. Angelo was a touch typist. Everyone, sculptor and investigators alike, were anxious to see how the sculpture and isolation experiment would turn out.

Hopes were high. But after only a few days in isolation the subject began pounding on the wall and yelling to be let out. The door was unlocked, and when he stopped hyperventilating (you know how artists are) was asked what the problem was. "You never gave me anything I asked for in my notes," said an enraged Mick. "But you only shoved blank pages out under the door," was the reply. The problem: After weeks of practice runs and training, and every major contingency planned for, someone had forgotten to put a ribbon in the typewriter!

I once experimented with myself to see what it would be like attempting to work in total darkness. One long, moonless winter's night, not being able to sleep, I went into the kitchen and attempted to make myself a cup of cappuccino without turning on the light. To insure total darkness I closed my eyes. My cappuccino maker is a very old, frustrating one, and although it is a machine, I feel it has a mind of its own—at least that is my excuse when the froth fizzles. In drawing the proper amount of water I used my finger to measure the depth in the beaker I use for this purpose (biologists have their own unique kitchen utensils). A dry finger determined the correct amount in the coffee measure. The hissing in my ears told me when the hot water had begun to spew out, and a scalded finger signaled steam blasting out of the frothing nozzle. The really difficult part was frothing the milk. My problem began with getting the milk out of the fridge without the appliance light turning on and spoiling the whole experiment. I solved that by unplugging the refrigerator. I found I used the different sounds of the raw steam bubbling the milk to create the perfect froth. I drank my creation in the dark (being pleased with the taste) and then turned on the kitchen lights to check for damage. I found froth on my nose meaning I had whipped up the milk to the proper tenacity and sticking power, and the stain on the front of my pj's had a nice, deep rich color. The only problem I noticed was that I hadn't fit the cover properly on the coffee grinder so a fine espresso dust had settled over the white counter. Also, the caffeine kept me awake the rest of the night. It wasn't until later the next day that my wife discovered I had left the refrigerator unplugged!

Now back to the conscription of volunteers for these kinds of experiments. Many come forward for idiosyncratic reasons such as overpowering curiosity as to just what it would be like. Others have practical reasons: Graduate students may want a quiet place to write a thesis, a way to escape an overbearing advisor, a rent-free place to live for a while, and a thousand other personal reasons. While these motives may all be sensible, life in an isolation chamber is not all fun and games: One must allow certain indignities such as providing periodic urine samples and sitting for six weeks impaled on a rectal thermometer.

In spite of all the difficulties associated with using human volunteers, there is a positive side also that is not available when lower animals are used: After the experiment is over each subject can verbally relate all his experiences (colored, of course, by his own prejudices) to his paymaster.

Another major category of subjects, which sometimes includes non-volunteers, comes from the hospital ward. Working with patients is very difficult: Many are sick, sometimes terminally. Just the opposite of isolation exists in this setting: Doctors, nurses, and medical students come and go at all hours of day and night. The lighting is by no means constant. And unfortunately, some experiments come to an untimely and abrupt end by the death of the subject (do not interpret that statement as meaning that death was caused by the experimentation). While this kind of research is difficult and often unrewarding, it is very important because eventually, as the learning curve is ascended, many lives will be improved and extended through a better understanding of human rhythms (this will be the subject of Chapter 3). This does not mean that animals, and, for that matter, even plants, cannot be good surrogates for humans; they can because we all have similar clocks.

## CAVEMEN AND THEIR CLOCKS: SLEEP/WAKE RHYTHMS

Over 30 years ago, a young speleologist decided that he could make a worthwhile contribution to science by descending into a 375-foot deep, cold (32°F) cave in the French Alps and living there alone for two months. He and his friends stocked the cave chamber in which he chose to live with a two-month supply of food, a tent with a battery-powered light, a field telephone, and a writing tablet. His watch was then taken and he was left in the cave to suffer alone. Day after day he sat there shivering in the cold (he claims his body temperature dropped nearly three degrees). Around him were constant small cave-ins and thick-as-molasses darkness. He amused himself by writing a best seller about his building libido (he was French), anxiety, and his few subterranean adventures (Siffre, M. 1964. *Beyond Time*. McGraw-Hill). Each time he ate, went to bed, and awoke, he called over a field telephone to his support group at the surface camp, where these times were recorded. Science never had a more devoted, self-motivated, self-appointed subject.

Our lonely caveman sat on his folding camp stool day after day, devoting some of this adventure time to making entries in his diary. He tried to keep track of the passing days, but on August 20, according to his log, his 37th day below ground, he was surprised when a rope ladder was lowered from above, and jubilant, champagne-bearing companions literally dropped in. They cheered, "You did it, you did it, today is September 14." To his great surprise, our protagonist had mentally lost 25 days . . . Whoever said time flies when you're having fun? He climbed to the surface where an exhilarated crowd awaited, cheering him as if something great had been accomplished, and even though it was midnight they rushed him off to one of those great French restaurants for escargot and wine and to read his diary (which, of course, was limited in scope to what a man can do for two months sitting alone on a camp stool in a dark, cold cave).

Although he had made his best effort at accurate record keeping, he had an insurmountable problem without a clock: Each time he awoke he had no way of knowing whether he had just briefly napped, or whether he had completed a full night's sleep. A major problem had been that he didn't count many of his short siestas, which unbeknownst to him were eight hours or more long!

This adventurer's truly interesting finding is seen in the chart made of his sleep/wake pattern constructed from the times of his telephone calls to the surface group.

Using the chart (see page 15), we learn that his living clock ignored his mental confusion and guided his body functions in a fairly orderly manner, measuring off days that averaged 24 hours and 30 minutes in length. To describe this change (everything in biology is named . . . even empty spaces, which are called *lacunae*) the word circadian (circa = about; diem = a day) was coined. It means that the living clock, when no longer forced into synchrony with, and by, the day/night cycle, may run slightly slower (or faster as the case may be). Thus, Siffre's cyclic behavior in the cave is described as a circadian sleep/wake rhythm. Tidal rhythms also may become circatidal in the lab. In fact it is very common for all types of rhythms to adopt a circa period when manhandled by experimenters.

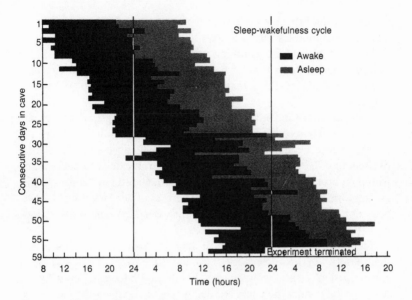

*Figure 2.1. The sleep/wake rhythms of a man living isolated in a cave. The dark bars represent the hours he slept each day; the lighter bars, his hours awake. Because his day lengths were longer than 24 hours the graph has been extended artificially by two days to the right so his rhythm can be shown without having it zigzag back and forth in a 24-hour wide plot.*

Siffre's period being longer than the natural day length, his rhythm remained out of synchrony with the day/night cycle at the cave entrance for the whole stay underground, except around day 40 when his day again corresponded to nature's outside the cave. Here his diary entry states, "For the last few days I have felt very optimistic, I suffer less from the cold; I am better adapted to conditions." There is no way to know if this observation was meaningful, or just a chance correlation. I'll guess the latter.

The above adventure appears to have started a competition to see who could stay underground the longest. The next contestant set a goal of 100 days, which he met and surpassed by five days. His cave's temperature held constant at 44°F, so for warmth, and because he was a romantic, he used only candles as a source of illumination. He had little interest in seeing what might happen to his rhythms so kept his wristwatch and intended to keep to his regular quotidian. But he soon found that he had problems trying to fall asleep at his accustomed time and commonly overslept in the morning. After three weeks he abandoned the timepiece and slept when he felt tired. Using a field telephone he had kept in daily contact with supporters outside of the cave, and using the timing of his calls it was found that he fell asleep on average 42 minutes later each day.

During this era of spelunking contests, the six best studies, which ranged in length from 8 to 25 weeks underground, produced circadian sleep/wakefulness rhythms that averaged 24 hours and 40 minutes in length.

These underground adventures are not without danger. A cavewoman, in setting a new record for length of time in isolation, became very depressed, according to her psychiatrist, when she finally emerged back into the light of day. Within the year she committed suicide. Another woman, who then broke the deceased woman's record, became so depressed during her time in the cave that she lost 20 percent of her body weight. I suspect that, as important as it is to keep one's weight down, this is a mode of dieting that will not become a fad.

Taking the story from Europe to America, two men decided to test their ability to trick their clocks into adopting a 28-hour period. They constructed makeshift living quarters in Mammoth Cave in Kentucky at a depth where the darkness was eternal, where the temperature varied less than one degree during the year, and in a section where tourists would not interrupt and gawk. They installed artificial lighting, beds, a table, and chairs. Crawling insects and other pests were a problem so they placed the legs of each bed, table, and chair in a bucket of water, producing protective moats around each that kept the experimenters from having surprise bed mates and uninvited dinner guests . . . at least those that could not fly. The two men forced themselves to live on a 28-hour day, with 19 hours of wakefulness alternating with 9 hours of sleep. The younger man was able to adopt this period in just one week, but the older man (Nathaniel Kleitman, pioneer sleep scientist who in 1953 discovered REM [Rapid Eye Movement] sleep and who lived to be 104 [suggesting that a daily nap can't hurt]) could not: He had great problems falling asleep unless the time to retire in the cave happened to coincide with his customary bedtime above ground.

These results are certainly interesting, but are they representative of humanity as a whole? Considering that the subjects discussed above are all courageous spelunkers, we should probably place them near the tails of the bell-shaped normalcy curve. But, I may be wrong in this assumption: As a university professor coddled in an ivory tower and mostly surrounded by meek-mannered colleagues bound together by a devotion to political correctness, how would I know which end is up outside the lab? Nevertheless, I will assume for now that more characteristic results must come from experiments in which standard creature-comforts are offered that will attract more everyman types of volunteers.

*Observations in Total Isolation*

Near the Max Planck Institute in Germany, some World War II bunkers were converted into constant-conditions suites for human habitation. Each apartment contains a small kitchen (a), a bathroom (b), and a combined living-, dining-, and bedroom (I & II).

Subjects enter through an antechamber (c) with an outside and inside door. The door's locks are specially designed so that when either door is open

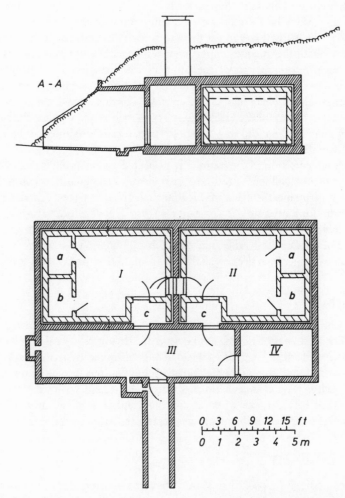

*Figure 2.2. Underground Bunkers*

the other is locked shut; thus, experimenter and subject can never meet accidentally in this foyer. In the antechamber is a refrigerator that is resupplied on a random time schedule with fresh food and an occasional bottle of the very satisfying Andechs beer. With scientific matter-of-factness, urine samples produced by the subjects are stored side-by-side with the food in the refrigerator, where they wait to be picked up by the handlers for lab analysis.

Watches and clocks were verboten, as were naps. After collecting certain samples for physiological studies, sometimes having to trail around the wires from a rectal thermistor probe, keeping a diary of their impressions, and often performing psychological and/or physical tests, the tenants could do whatever they pleased during two- to six-week stints in isolation. Many incarcerates were students who used the time to prepare for exams (finally, a way to get them to study). The only way of communicating with the outside support crew was to leave and receive letters in the refrigerator (remember the problem Mick Angelo had). Irony of ironies, the last words of those entering isolation (remember they were mostly students) were logical and upbeat, "The first thing I am going to do is catch up on missed sleep." Contrary to their stated craving, the first thing many of them did was stay up for unnaturally long intervals before finally settling into a fairly regular routine.

Movements around the room were measured automatically by depressible plates on the floor. A pressure plate under the bed signaled sack time. In most experiments the lights were left on at all times, but a few experiments were run for a brief time in constant darkness. The sleep/wake rhythms of 147 subjects were studied in these bunkers; the average period of the group was 25.2 hours, with a range between 17 and 38 hours. This average is significantly longer than sleep/wake rhythms produced in the caveman studies.

### Neonatal rhythms

The initial suggestions of the presence of a sleep/wake rhythm emerge in our first week of post-uterine life; however, the rhythm is so weak that it can be discerned only by measuring the total minutes awake versus those asleep. One study found that in the first postpartum week, the average accumulation of sleep was 6.5 hours between 8 A.M. and 8 P.M., and 8.5 hours between 8 P.M. and 8 A.M. By ten weeks of age, the daytime sleep had dropped to about 3.5 hours, while the nighttime average stretched to ten hours.

### A Womb with a View

Other studies have peeked right into the uterus—using ultrasound—and monitored unborns' movements, heart beat, and even breathing.[2] Daily rhythms in movement have been found to first appear between weeks 24 and 30 of gestation. While rooming in the womb the fetus is exposed to all its mother's rhythms, so no conclusion can be reached yet about the presence of a fetal clock.

While on the subject of fetal life, did you know that in the later stages of the fetus's uterine development it can hear? This bit of knowledge comes to us from psychologists. Using a Pavlovian protocol, an investigator would ring a bell and punch a pregnant woman in the belly. The intrauterine occupant would respond to the blow by flipping around a bit. After many repetitions of hitting below and above the belt, the investigator/pugilist needed only to ring the bell and the fetus would begin to duck and feign. Years ago my pregnant wife and I made a similar observation, but in a much more civilized way, while we watched the classic movie *Exodus*. After listening to several repetitions of the show's powerful musical theme, my unborn son did something akin to a fetal funky chicken each time it played. She and I felt her tummy and whispered so many times during that show that a man sitting next to us whispered, "You should get her to a hospital before she is dilated to ten centimeters!" In case you are wondering, my son does not remember ever hearing the theme from *Exodus*.

## Battling Clocks; The Rhythm of a Blind Subject

One way of overcoming the reticence of people to subjecting themselves for study in constant darkness is to use blind subjects—they have no idea whether the lights are on or off. In one such approach a 28-year-old male university student, blind since birth, asked the university health services for help. He reported that at two- to three-week intervals he tended to fall asleep in class and feared he would fail his statistics course. Between these daytime soporific episodes, he slept very well and only at night and was alert during classes. It was decided that his sleep/wake pattern should be monitored both at home and in the hospital. The results of the study are seen in the figure on page 20.

We see in A that each night he fell asleep closer and closer to dawn. By day ten, he fell asleep at 5 A.M. and awoke at 1 P.M.; on the next day he was admitted to the hospital where he was allowed to sleep any time he wished. Free of worry (his statistics teacher, a mathematician, and therefore conventionally demanding, was very understanding in this case), he followed these directions, and as you can see during his stay in the hospital (bracketed by B) the nurses actually let him sleep. Each day he fell asleep 48 minutes later on average.

After 26 days in the hospital he was sent home with instructions to force himself to sleep at night only. But as you can see in the figure, even

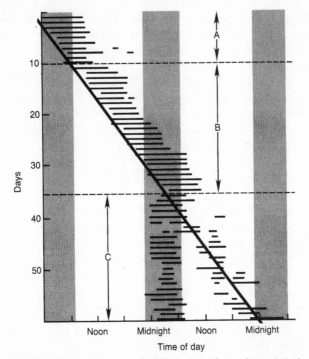

*Figure 2.3. The sleep/wake pattern of a blind subject first at home (A), then in the hospital where he was allowed to sleep whenever he wished (B), and finally back home at nights and in the classroom during school hours (C). The bars represent the intervals that he slept.*

though he made a heroic effort, and did get some shut-eye at night, he still tended to fall asleep during daytime classes (the days bracketed by C). The line I have extrapolated from the times of sleep onset during the first 36 days of observation tends to represent the trend of the sleep intervals. This unfortunate lad appeared to be in a lifetime struggle with his living clock: He wanted to exist on a 24-hour basis and be in sync with the rest of humanity, while his clock was equally obsessed with guiding him along at a 24.8-hour period. The hospital stay defined the problem, but did nothing to alleviate it.

### RUNNING HOT AND COLD: THE BODY-TEMPERATURE RHYTHM

In the old days, long before electronic oral thermometers, the instrument was made of glass with an etched-in red arrow pointing to the 98.6°F reading. That value was considered normal oral temperature and often determined whether as a child you could stay home from school. The fact that the mercury was seldom right on the arrow, but was only close to it, was Okay because 98.6°F is not really normal, and one's temperature changes in a rhythmic fashion throughout the day. This rhythm was first described by a physician in the British army in 1845, who used himself as the experimental subject.

Temperature taking in those days in England was only roughly the same as it is today. For one thing, the thermometer was over a foot long, and the bulb containing the mercury was mouthful in size. Dr. Davy had bent his thermometer into a right angle at the base, so while he held the bulb in his mouth the tube stuck straight up in front of his nose, where, because he was cross-eyed, he could easily read it. His protocol was to first keep his lips tightly closed for 15 minutes for his mouth to warm up, then somehow insert the enormous bulb under his tongue, and hold it there for several more minutes until the generous mercury reservoir equilibrated with tongue temperature. He made these measurements almost hourly during waking hours for eight months, leaving little time to say, "Take one of these pills and come back tomorrow." He found that his temperature was lowest in the wee hours of the morning and rose to a high in the late afternoon/early evening.

In 1868, Carl Wunderlich conducted a follow-up study in which he took one million temperature readings from 25,000 people. His study differed in that he used armpit temperatures because his thermometers were even more cumbersome than Dr. Davy's. Each reading took about 18 minutes (meaning he gathered 300,000 hours of armpit-temperature data!). He confirmed that human body temperature undergoes a daily rhythm, and it was he who gave us the 98.6°F value as the daily average.

But 98.6°F is wrong. Modern studies tell us that the average, daily, oral temperature for healthy 18- through 40-year-olds is 98.2°F., with women being warmer, 98.4°F., and men colder, 98.1°F. The bottom line here is that for the last 150 years, thermometer makers have been misleading—with their little red arrow—the entire medical profession and Dr. Mom.

Dr. Davy's discovery of the temperature rhythm has since been confirmed many times; the following is an example of what an average curve looks like.

The temperature curve peaked from 7 to 8 P.M. and was lowest sometime around 5 A.M. It is interesting to note also that Englishmen used in this measurement really are colder than Americans: Their average daily temperature was 97.6°F. (And I thought it was just the beautiful way they speak that made them so cool.)

Your critical minds in full gear, you must be thinking that such a curve is only a consequence of being flat on one's back in bed at night, versus

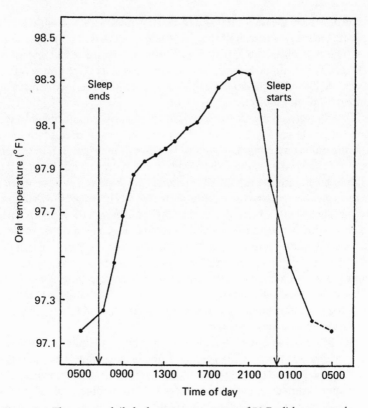

*Figure 2.4. The average daily body temperature curve of 70 English seamen who took their temperatures every hour during the day, and at 2-hour intervals at night.*

awake and active during the day. That view, endorsed by many human physiologists, is, of course, largely correct: The daily alteration in activity does play an important part in the genesis of that rhythm, but never underestimate the living clock's fundamental contribution to that rhythm and many others (soon to be discussed). To avoid activity as a contaminating factor, subjects were simply kept in bed all day, and to eliminate the influence of meals (the digestion of food releases heat), the volunteers were either made to fast, or were given identical small meals at short intervals throughout the day and night. Their temperature rhythms persisted. Daily temperature measurement of a tragically paralyzed polio victim were made over 16 consecutive months, and it was found that his temperature rhythm persisted the whole time. When one is sick with, say, the flu, the fever produced does not erase the temperature rhythm, it just increases its amplitude; thus in the late afternoon and early evening one suffers the most, while in the early morning the infirmed's temperature can be close to normal.

The temperature rhythm will persist when activity and eating are controlled. The other variable, the day/night alternation in light, has also been eliminated by studying subjects in the underground bunkers in constant illumination. Their temperature rhythms persisted quite nicely.

Two well-known results of sleep deprivation are a continuous reduction in body temperature and an increase in fatigue. In a four-day study where the subjects were kept awake, the body temperature decreased gradually, but the temperature rhythm persisted, for example, the daily peaks and troughs of the rhythm decreased slightly each day. Using a subjective test of fatigue (the repeated question: "On a scale of one to five, how tired are you?") given during sleep deprivation, it was found that fatigue increased daily, but it did so rhythmically, with peak times coming in the late night when the body temperature is low, followed by a noticeable partial recovery after dawn. But we all knew this already: When we were students either pulling an all-nighter to cram for a test, or during marathon partying, around 3 to 4 A.M. one feels pretty awful, but then as dawn arrives we get a second wind.

As an aside, do people deprived of sleep make up for that loss when they do finally get to sleep? Not really. For instance, a 17-year-old boy attempting to set a new record stayed awake for 11 days (264 hours). Then when he finally slept the first bout lasted only 14 hours, and after that resuscitation, he slept only 8 hours per night thereafter.

*Human Birds*

It is possible to identify some avian characteristics in about 40 percent of the human population. Some are early birds (larks), people who arise at the crack of dawn, or earlier, and jump out of bed aflutter and raring to go. The nest mates they leave behind may be the night owls, a sluggish-in-the-morning group that have great trouble waking up, getting out of bed, finding the kitchen, and making that first pot of coffee. The tables are turned at night: The early birds poop out and leave the party early and retire to their nests about the time the night owls are just getting into high gear, and among other activities, grousing about the departing early birds. For every one early-bird male, there are three night-owl men. The ratio is one to two for the ladies. The rest of us seem to be sort of intermediate, chimera-like, tortoise/wallflower/practical types.

These two birds, easily identified in the field, can also be recognized by certain personality tests that measure sociability. Forty-seven English seamen were identified as either introverts (early birds) or extroverts (night owls) in this manner. Their oral temperatures were taken periodically for two days and the results averaged. The mean daily temperature of both groups was 97.9°F, and, of course, both groups' tempurature was rhythmic. But the rhythm of the introverts indicated that they warm up sooner in the day, peak sooner, and fade sooner than the extroverts.

This same group of subjects was further tested. This time it was their efficiency that was measured, and the tests were not what you might think would be of interest to the Queen's Navy. The men were not tested as to how accurately they fired the guns, how well they could navigate, their ability to use radar, et cetera. Instead, five times a day they were required to cross out as many letter E's in *Punch* magazine as they could in 30 minutes. The reason for the choice of this landlubbing task was that it could be easily quantified. Each time they performed this no-brainer exercise their temperatures were taken. It was found that both groups were least efficient at 8 A.M., but that the early-birds did better than the night owls; both groups did best at 8:30 P.M., although the night owls did much better than the early birds at this time. As you will see as we proceed, the peak of the temperature rhythm often corresponds to our peak efficiency in many tasks.

The same sort of measurements have been made on students taking psychology courses. These captive subjects are given a variety of chores that are repeated at given intervals throughout the day. Each time a task is undertaken, the students take their temperatures. The tasks are rather inane, but have been chosen because they are easily quantifiable. The students see how fast they can separate a deck of cards by suit, how fast they can multiply two large numbers and how accurate their calculations are, how quickly they responded to a light signal, the time required to pound different-sized pegs into their corresponding holes in a board, their vigilance in monitoring a simulated radar screen, and the like. All of the results are plotted against the time of day and the oral temperature at which each task was performed. It is found that the subjects' mental and physical skills tend to mimic the forms of their daily temperature rhythms, meaning that most of them do best in the late afternoon and early evening when their temperatures peak for the day.

Athletes should pay close attention to what time of day they execute best. Careful studies of certain sports have defined the best performance times. Just five events are listed here as examples:

| | |
|---|---|
| Running | 7 P.M. |
| Shot Put | 5 P.M. |
| Rowing | 5 P.M. |
| Grip Strength | 8 P.M. |
| Swimming | 8 P.M. |

When a coach gives his team new plays to memorize, the players retain them much better if the instruction is given after 3 P.M. than in the morning. It would be easy to jump to the conclusion that there is a cause-and-effect relationship between temperature and performance. While that relationship may hold in some cases, most likely in tasks that involve physical strength, many of our mental attributes do not follow our daily temperature curve.

In about half of experiments that study efficiency as a function of time of day a pronounced, temporary, depression in the efficiency curve occurs after lunchtime; it is referred to as the postprandial dip. Although there are some students who are sure that the dip is a result of bad lunches served in the school cafeteria, this does not have to be true: The dip can occur even if lunch is skipped. Countries in which siestas are part of custom may well be practicing a ritual with a sturdy biological basis.

### EACH CLOCK TO ITS OWN:
### INTERNAL DESYNCHRONIZATION OF RHYTHMS

Each of the body's many rhythms holds a particular phase relationship with every other rhythm, meaning that all the rhythms do not peak at the same time, but hold constant intervals between each other. When a person is subjected to a stint in constant conditions, his rhythms usually adopt a period slightly longer or shorter than 24 hours in length, and not everyone of them assume the same length. I'll give you two examples, each an extreme response to life in constant conditions. The record below is of a young man during 25 days in constant light.

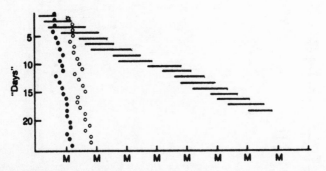

*Figure 2.5. The daily sleep intervals (horizontal bars), and the peaks (solid dots) and minimas (open circles) of the daily body temperature rhythm of a man in constant conditions. Twenty-four hour intervals are indicated on the abscissa where M represents midnight. The "Days" on the ordinate refer not to real 24-hours days, but only to the number of intervals the subject slept.*

You see that the subject retired and arose considerably later each day (the average interval between one snooze and the next was 33.6 hours—an unusually large deviation from a normal day length). His temperature rhythm assumed an average period of only 24.6 hours. The sleep/wake rhythm and the temperature rhythm desynchronized from one another in a dramatic way. During this stint in the bunker had the young man tried to estimate how long he had been there, using his sleep/wake repeats he would have guessed 18 days, but counting the peaks of his temperature rhythm he would have chanced about 24 days.

The other example is even more dramatic. In this case the subject was a young woman whose sleep/wake rhythm first shortened to an unprecedented average of only 12.5 hours!

During each eight-hour interval while she was awake she ate three hearty meals! Then she slept on average 4.3 hours (one would guess it was not hunger that awakened her). But after 16 of her truncated days and even shorter nights, her sleep/wake cycle spontaneously converted to a more normal 24.5-hour period. The diary she kept indicated that she was totally unaware that she was living mini-days and eating full meals roughly every four hours, nor was she cognizant of the sudden period increase to 24.5 hours. While all this was going on her temperature rhythm was just slightly longer than 24 hours.

Both of these cases are representations of internal desynchronization. That separate rhythms in our bodies can have different periods in con-

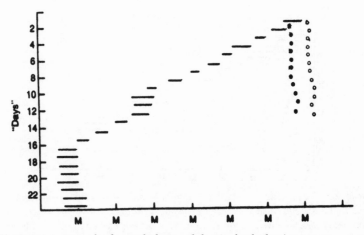

Figure 2.6. An example of a much shortened sleep/wake rhythm in constant conditions. The symbols are the same as in Figure 2.5.

stant conditions is most often interpreted as indicating that different rhythms are governed by specific clocks that each runs at their own speed. Also, as we will see in chapter 4, internal desynchronization can be an important contributor to the jet-lag symptoms we experience after long-distance rapid east or west travel across the face of the earth.

## EXCRETORY RHYTHMS: BEDWETTERS ARE THE EXCEPTION

Important substances are distributed throughout our bodies dissolved in our blood. Some of these substances must be maintained within a rather narrow range of concentrations, otherwise serious consequences (one of which is death) are inevitable. It is the kidneys that perform much of this regulation. If, for example, you eat a very salty meal, large amounts of sodium and chloride end up in blood; it is the kidneys' job to excrete the excess. At the other extreme, during a long jog on a hot summer day a large amount of salt could be lost as you sweat, so the kidneys must conserve it or you may faint. The regulatory process is a very important one requiring a great deal of energy: The kidneys consume more oxygen in their labors that does the heart.

That the kidneys' output was rhythmic was discovered way back in 1843, when it was observed that—with the possible exception of bed wetters—the amount of urine passed was less at night than during the day. No particular importance was attached to the discovery, it was just thought to be due to the fact that people do not drink or eat while sleeping. However, after the turn of the century it was found that the rhythm in urine volume persisted even when people were kept in bed all day and made to fast, or were fed identical small meals at short intervals all through the day and night. When the various ionic components of the urine (such as sodium, chloride, potassium, phosphate, et cetera.), and certain hormones were examined they were also found to vary in concentration in a rhythmic manner and under clock control. Doctors order a urinalysis during every physical exam in an attempt to identify things that are going on internally without having to go inside our bodies and look around. Full knowledge of the rhythmic alternations therein can play an important role in a doctor's interpretation of the meaning of the measurements and the state of one's health.

Our kidneys receive commands from other parts of our body via hormones circulated in the blood and nerve impulses coming from the spinal cord. But there is also a control center within a kidney: Each has its own living clock that can be seen in action when a kidney is removed and transplanted into another person!

In a study of 25 people who received a kidney transplant, in seven of them the excretory rhythms persisted unchanged in form and phase from that displayed when the kidneys were still residing in their original owners. This transfer eliminated a nervous-system input because in its new location the kidney is not connected to the recipient's nervous system. However, the relocated kidney does receive hormones from the recipient's blood, so an endocrine control is not lost. But, while the other 18 transplanted kidneys also continued to be rhythmic, their phase had mysteriously become reversed: When the rhythms of the recipient's remaining kidney peaked, the rhythms of the borrowed kidney were at their daily low points. This even happened with a kidney transplanted to an identical twin! And these reversed rhythms—whose cause is unknown—remained transposed for as long as a year. Thus, it is not necessarily the recipients' nervous system or hormones that are providing any timing information; each kidney has its own clock.

Surgeons were curious whether the time of day that a kidney transplant was done played any role in the success of the outcome. Rats, which are often substituted as pinch hitters for humans, were used to investigate that question. Two different strains were employed, specially chosen because there was little histocompatibility between them; this meant that a transplanted organ from one rat to another would soon be rejected. Both kidneys were removed from a rat, and a kidney implanted from the other strain. The time at which it was rejected was signaled by the death of the recipient rat.

All the rats were kept in natural conditions throughtout the day and night, meaning that any one individual's rhythm should be in sync with all the others'. Transplant operations were done at 4 A.M., 8 A.M., noon, 4 P.M., 8 P.M., and midnight. All the rats who received their kidneys during the hours of daylight rejected them in less than four days; while those who received theirs during darkness lived much longer, in fact half of those who got their new kidneys at 8 P.M. lived longer than three weeks! Thus, time of day plays a big part in the success, or lack thereof, in kidney transplantation operations—at least it does in rats.

Rats are nocturnally active, while humans are diurnally active; in other words, the phases of their rhythms are reversed. If one wished to extrapolate results from experiments on rodents to humans, the most successful outcomes for human-to-human kidney transplants should result after operations performed at 8 A.M. (if done by an early-bird surgeon). Any readers want to volunteer for a late evening transplantation to test that speculation?

## The Heartbeat Rhythm: Space Shots & Transplants

The rhythm I'll discuss here is not the rate-per-minute heartbeat; it is an overlying rhythmic change in the rate that takes place during the interval of a day. The lub/dub frequency is easily modified by a variety of exogenous and endogenous stresses: Eating a meal causes a rate increase, as does physical activity; so does the emotional stress that may occur when a doctor measures your blood pressure (the increase attributed to the white-coat syndrome). Emotion combined with activity, such as during sexual intercourse (depending, of course, on how acrobatic a couple wants to be in the bedroom) can really speed up the rate (in rare cases the combination can cause the heart to stop—this is called the Garfield Syndrome after a well-known Hollywood actor who left this world *en copula*). On the other hand, inactivity and rest reduce the number of heartbeats per minute. And embedded in all these responses is the activity of the body's clock that speeds up heart rate during the daytime and reduces it at night.

Heart rate is also a direct function of body temperature, the warmer one is the faster the heart beats. Before the invention of the oral thermometer, doctors first determined if a person had a fever by placing his cold hand on the customer's forehead, and then moved it to a superficial artery to quantify febrility by counting heartbeats: every 10–15 beat increase above average signifies a 1°F. elevation in body temperature.

With this background it becomes obvious that to examine the role of the body's clock in controlling heartbeat rates, several other physiological parameters must be held constant, or at least be compensated for. Here again then, subjects are kept in bed, starved or fed identical small meals at regular short intervals, and, of course, are prohibited from having sex. Couch potatoes make good subjects. When these conditions are met, a clear daily heart-rate rhythm can be demonstrated.

This rhythm will even persist during the weightlessness of space flight in an astronaut crammed into a space capsule. The study I will describe was done with one of the nation's superheroes, Frank Borman, as he piloted the Gemini VII space craft around the earth 192 times during 14 days in space. His heart-rate rhythm persisted throughout his dizzying flight with a period of 23.5 hours and was only temporarily disrupted at blast off and again at splash down. Needless to say, these two times were quite emotional even for a nerves-of-steel astronaut: At liftoff his heart rate increased about 50 percent, and during his high dive into the Pacific Ocean on his return to Earth it jumped to 200 percent of normal. Commander Borman had found an activity even more heart stimulating than sex! Bungee jumping probably runs a close second.

Coordinating the beats of our heart's individual cells is an internal control center called a pacemaker. When a diseased heart is removed, a flap of heart tissue, that happens to contain the pacemaker is left behind as a place to anchor the replacement heart when it is introduced. The new heart comes with its own pacemaker that continues to govern its beat. After a successful operation, when electrocardiograms are made the activity of both the old and new pacemakers appear on the record. Curiously, in the first operation in which this was followed, the two pacemakers' output rhythms, while both having the same period length, were unexplainably out of phase with one another; one peaked 130 minutes later than the other. This is interpreted as indicating that the human heart has its own clock, just as was demonstrated for the human kidneys.

One million Americans have a heart attack each year. Here is some information that is worth knowing, especially if you are over 50.

Two kinds of enzymes involved in blood clotting are present in our circulatory system. They are tissue plasminogen activator (abbreviated as tPA) and plasminogen activator inhibitor type 1 (also known as PAI-1). Everyday holes form accidentally in our blood vessels that are quickly plugged with a clot. Once these wounds have healed sufficiently, the clot must be removed so that it does not restrict the flow of blood through the vessel. Within a clot is the means of its own ruination, a molecule called plasminogen. If it is activated to plasmin, the clot self-destructs. It is the tPA that activates the conversion; the sequence is as follows:

```
                      tPA
    plasminogen ————————————> plasmin + "gen"
                                  ↓
                  blood Clot ————————> clot destroyed
```

Even though the heart is filled with blood, that blood does not nourish it and carry oxygen to it. Instead, the heart muscle's blood supply is delivered via branches of the coronary artery, and when a clot forms in one of them the blood becomes dammed up behind it, the part of the heart it serves starves, causing a coronary heart attack. A doctor can attempt to save such a victim's life by injecting synthetic tPA directly into his or her circulatory system, in which, after it travels to the coronary artery, it causes the breakdown of the clot and restores circulation to the heart again. It's a neat clot-busting trick that works well if administered soon after an attack begins.

But remember, tPA, which occurs naturally in the blood, is present in too large amounts it would break down those many accidental-wound clots as soon as they form and we could bleed to death internally. That is where PAI-1 plays a role: It inhibits the action of tPA. It should be obvious that a delicate balance must be maintained between the concentration of the two substances in the blood if we are to reach old age.

As it turns out when measurements are made, the amounts of both tPA and PAI-1 vary rhythmically over the day; tPA is lowest at night and the early morning, about the times that PAI-1 is highest. Thus, from 3 A.M. on to about 9 A.M., there isn't much tPA circulating to activate plasmin, and further, what is circulating is greatly inhibited by an abundance of PAI-1; this condition sets the stage for a possible heart attack.

The story gets even worse. Another molecule, fibrinogen, also circulates in our blood, and the body's clock makes it most abundant during these early-morning hours. The higher the concentration of circulating fibrinogen the greater the chance of blood clotting, because given a variety of different stimuli, fibrinogen breaks down into fibrin, a well-named substance that forms a physical net of fibers across a wound; the net is a precursor of a clot.

Blood platelets are tiny, membrane-covered dollops of cytoplasm. They contain a substance that initiates a chain of chemical reactions that results in the production of fibrin, and thus a clot. Their delicate membranes rupture easily, spilling out the clot initiator. The platelets are very abundant (something like 300,000 in each drop of blood) and the number circulating at any one time is under the control of the living clock; unfortunately they, along with the amount of fibrinogen, are most abundant in the early morning. Presented with all this information you will not be surprised to learn that most heart attacks occur in the early morning, and that is the time when most of us die from heart attacks.

Aspirin, circulating in the blood, reduces the ability of the blood platelets to break open. Thus, aspirin (but not acetaminophen) prevents plugging the coronary artery and thus heart attacks. For those of you who have been instructed by your doctor to take an aspirin every day (or every other day), the best time of day to take one is on awakening, when the platelet and fibrinogen rhythms are at their peaks.[3]

## CELL-DIVISION RHYTHM

Daily rhythms in cell division have been known since 1851, when they were first described in the root and stem tips of flowering plants. Such rhythms are also common in single-cell algae, protozoa, and even small mammals. Could they also exist in humans?

The answer to that is difficult to obtain because even the easiest procedure requires taking punch biopsies from people's skin. Most people fear getting shots and are terrified even at the thought of having a series of skin plugs plucked out of their bodies at hourly intervals. The project was first undertaken in a very clever fashion, using skin that was to be discarded anyway. A special advantage of this particular study was that all of the volunteers were living together under the same conditions and were within a day or two of the same age. When I presented this story to a class of students, one sharp and circumspect woman asked if the subjects' sex also had been taken into account. I was able to guarantee that all the subjects were male, for it was their discarded foreskins that were being studied. What I had to admit was that none of the subjects had given informed consent since they had not learned to read or write by the end of their first week of life.

The 1939 study was carried out in a hospital nursery in New York City by physician Zola Cooper. With considerably more finesse that Lorena Bobbitt,[4] she snipped off foreskins at most all hours of the day and counted the number of dividing cells in each sample. Then, when she plotted out her data against the hours of the day she found an impressive daily rhythm that peaked about 9 P.M.; this cycle is a true classic . . . the first ever described daily rhythm in penis growth!

## MISCELLANEOUS RHYTHMS

*The Highs (and Lows) of Alcohol Metabolism; don't have one for the road.*

Many of us look forward to cocktail hour after a day's toil. We enjoy the taste and the nerve-soothing effects a drink provides. The alcohol begins to be absorbed, unchanged, as soon as it enters the stomach, in which it passes into our blood and is circulated throughout the body. When it reaches the head it marinates our brain cells producing the desired effects, such as gentle sedation, intoxication, or unconsciousness . . . depending on a drinker's goal or carelessness. One stays under its influence as long as the alcohol remains circulating in the blood.

The body's liver assumes a different attitude about this preprandial happy hour: It recognizes alcohol as undesirable and rids the body of it by breaking it down to harmless carbon dioxide and water. The key enzyme in this degradation process is called alcohol dehydrogenase. This enzyme is also found in the lining of our stomachs where it plays the same role. With that information we are armed with yet another way to distinguish men from women. You may have noticed that women get tipsier more rapidly than men on the same amount of alcohol. Conventional wisdom explains this by noting that women are smaller than men, but cocktail volumes are all the same size. As logical as this may seem, the axiom is wrong, as can be demonstrated in substituting a hypodermic syringe for a stemmed crystal martini glass and injecting equal amounts of alcohol directly into the circulatory system. This presentation, which bypasses the stomach, knocks both sexes equally silly in the same amount of time. How can this be? Remember that I stated above that alcohol dehydrogenase is the key player in the breakdown of alcohol, and that this enzyme is also present in the stomach lining. Men have much more in their lining than women, thus much of each drink they consume is broken down in their stomachs before it can be absorbed, and therefore never even reaches the brain. The end result is obvious: Women get more bang for their happy-hour buck!

As already stated, the high we experience lasts as long as the alcohol remains as such in our bodies. And how long it lingers depends to a large extent on—of all things—our living clock. When the experiment that elucidated this was planned, eager, thirsty volunteers lined up by the score outside the lab/bar. I feel I must mention that most were budding biology students who claimed that they chose to offer their services so as to receive the HDL-building benefits of a drink or two (HDL, high-density lipoproteins, remember, is the good cholesterol, if one wishes to assign morality to a molecule). The lucky few volunteers selected were treated with this kind of party: Each was given one ounce of whiskey to drink at the top of each hour, and 59 minutes later blood was drawn and the amount of alcohol in it determined. This was done at hourly intervals throughout most of the day, but less frequently between midnight and 5 A.M. The resulting alcohol-metabolism curve is shown in Figure 2.7.

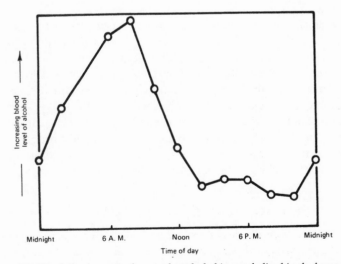

Figure 2.7. The daily change in the rate that alcohol is metabolized in the human body.

The following conclusions can be drawn from the curve. Between 10 P.M. and 8 A.M. the liver and stomach work slowly at breaking down alcohol, so that the amount left circulating at the end of each hour steadily increased. If one on a limited budget has a singular goal—to get as blotto as possible—this is the time to do it! But after 8 A.M. the liver gets into high gear and not only breaks down the alcohol in the last drink, it also begins to destroy that left over from previous drinks. By 2 P.M. it has rid the body of its previous accumulation of hooch and metabolizes the new incoming load as fast as it is consumed. This continues right through the cocktail hour until about 10 P.M. when the liver again becomes less effective. Thus, the "one more for the road" after the bewitching hour can land a party animal in a cage—or worse if driving.

### When, Instead of Where, Does it Hurt?

Pain is a rather difficult term to define. For instance, suppose when a small child walks by its parent, the adult leans over and swats him on the bum. If the child looks up and sees anger on the father's face, the swat feels painful, and the child may burst into tears. But if the parent is smiling the swat is termed a "love pat" and brings forth only childish giggles. A hospital patient suffering pain sometimes gets relief from an injection of sterile water if he is told that the shot is morphine. Then there are those who enjoy pain! Now I'll add another dimension to the confusion: Humans' interpretation of the degree of pain is under the control of their living clock.

Legend has it that a seemingly sadistic dentist, constructed a small version of a cattle prod and used it to deliver an electric shock to peoples' teeth. (There is little need to mention that many fewer volunteered for this study than did for the alcohol-metabolism experiment.) Every two hours throughout the day the same intensity shock was administered, and the electrocuted recipient evaluated how painful it felt. In spite of the fact that the stimulus was always the same intensity, the recipients interpreted it as increasingly more painful throughout the day up to a peak at 6 P.M.; after that time the agony lessened. As luck would have it, it turns out that normal business hours proved to be the most painful times.

### From the Beginning to the End, the Clock is with Us

In some situations the clock functions as a gate. Here is a barnyard example of a gated response. A cattle pen is filled with cows, all of which would like to get out and eat the grass that is, of course, greener on the other side of the fence. Farmer Brown opens the gate and out they go. Let's say

Farmer Brown fears being trampled as the cattle rush out, so he attaches a clock to the gate latch that will open it in five minutes, giving him time to get out of the way. That would be a clock-gated situation.

Human birth and death times (clearly one-time events) are considered gated responses, for example, your body gradually ages and then . . . kaput, the Grim Reaper's gate opens. But is that gate clock operated? In search of an answer, 433,000 hospital death certificates were inspected. It was found that death comes most often in the wee hours, peaking at 5 A.M. Again, conventional wisdom has it wrong: These data suggest that it is our biological clock, not our spouse, that has the last word. There is great wisdom in this quote from Russell Baker from *The Sayings of Poor Russell*: "The dirty work at political conventions is almost always done in the grim hours between midnight and dawn. Hangmen and politicians work best when the human spirit is at its lowest ebb."

Now let us switch to birth times. In the modern study that I will report here, the first thing that was learned was that one could not use the birth records from private hospitals because births tended to be artificially induced on Fridays—the snide interpretation of which is so as to leave weekends free for doctors to play golf. Thus, data from public hospitals had to be used; the birth times of 2,082,453 spontaneous live deliveries were collected. Like death, the birth-time curve found takes the form of a daily rhythm, with the peak coming close to 4 A.M. (no wonder obstetricians often appear so haggard). Thus, a baby grows in its mother's womb until it is sufficiently mature to be born, but then there is a tendency to delay birth until the maternal clock permits the baby's emergence.

## I Tell Time; Therefore I Am

Lastly, here is a true story about a man who literally became a clock in death. He was cremated and his ashes placed in a three-minute egg timer for his wife's use in the kitchen. (From the *Otago Daily Times,* in Dunedin, New Zealand, July 15, 1997.)

## Biorhythm Bunkum

Before leaving this chapter let me ask a favor of you. Never refer to biological rhythms as "biorhythms." The former belong purely to science, the latter are pure balderdash. Somehow, almost the entire American public was sold a bill of goods throughout the 1970s and early 1980s on the biorhythm method of predicting—just by knowing the date of their birth—for the rest of their lives, all of the forthcoming times of high and low emotional states, dangerous times, times when you will be strong or weak, when to take risks, et cetera. Biorhythm dogma proffers the reality of a family of three long-period rhythms: 23, 28 and 33 days in length (none of which actually exist). Supposedly, how the rhythms combine in different phase relationships determines what may happen to you on that day: One might win the lottery, marry Elizabeth Taylor or Pierce Brosnan, or even die.

One year the biorhythms of all the players in the superbowl were calculated and commingled by team in an attempt to predict where to place one's bet. Anyone can calculate his or her own biorhythms with just pencil and paper, but hand-held calculators and computer programs were sold by the thousands to the naive—as were many books on the subject. P. T. Barnum was right. The con men foisting this method of haruspication on the gullible public gained credence by claiming it was based on scientific fact, and to reinforce this canard, they gave it a name of a respectable field of biology. Please believe me when I say there was nothing scientific about biorhythms. Happily, most of the nation finally figured that out for themselves. But some reporters, and others who should also know better, still use the term when they really mean biological rhythms. Using "chronobiology" is the clearest way of avoiding the *faux pas*.

That said about the vulgarization of the term "biological rhythms," let's turn to a common misuse by the media of "biological clock." For the last several years reporters have chosen to ascribe to our living clock the governance of how long we live. Here is a recent example taken from an article entitled "The Cells of Immortality," that appeared in *U.S. News & World Report* (March, 2000):

> The discovery of biological clocks ticking away in each of your cells, and a knowledge of how to reset those clocks, open the possibility that a human would never die—at least not from old age.

Poppycock. While fatalism has always been a popular philosophical approach to death ("When your time is up, it's up"), there is no evidence that death is caused by the last grain of sand falling in a living hourglass. The clock in each cell that we know about measures out intervals of the tide, the day, a fortnight, a month, and a year . . . only. Our clock may determine the *time of day* we die, but not our longevity. This biological-clock maundering of the media just represents careless reporters romancing science.

T H R E E

# RHYTHMIC PHARMACOLOGY

The fundamental rate at which chemical reactions proceed is a function of temperature: As heat is applied, chemical changes take place more rapidly. Heat increases the kinetic energy of atoms and molecules, causing them to collide more frequently, permitting them to react with one another. An obvious example is the time required to turn liquid pancake batter into a solid pancake; the conversion depends on how hot the griddle is. Looking at it the other way, we keep food in the refrigerator to slow down the reactions that cause rancidity and spoilage.

Our lives depend on the myriad chemical reactions that take place within our tissues and organs, beginning with the digestion of meals we consume and moving to the conversion of the digestive products into human tissue. The warmer our deep-body temperature, the faster our internal chemistry churns; and as discussed in the previous chapter how warm we are at any time of day is in part dictated by our living clocks. Thus, because of our temperature rhythms, other factors being constant, dinners are digested faster than breakfasts because our temperature rhythm peaks around dinner time. The temperature rhythm also contributes somewhat to the additional prowess we exhibit in athletic events that are scheduled in the late afternoon and early evening; the change is produced in part by the chemistry of muscle contraction being speeded up at the temperature-rhythms maximum.

When we are sick and feverish, our temperature rhythm persists but in an augmented form. We are most febrile in the late afternoon and the early evening, during which we suffer a bit more. But our temperature returns closer to normal at dawn. One slight benefit of having a fever, if one is overweight, is that calories are burned off faster (but it is a masochist's way of dieting).

Now think of what happens, not to you, but to the medicine you take. Medications are recognized by the body as something foreign to it and, therefore, something to get rid of. Thus, beginning soon after you pop a pill, its beneficial ingredients are attacked and eventually broken down into harmless substances. It follows that because the body temperature peaks in the early evening it is inevitable that the destruction of medications is more rapid at that time. This is one of several reasons that the effectiveness of medications differ as a function of the time of day that they are taken. Another reason is that most physiological events in the human body are rhythmic: Here is a list of a few of them

| | |
|---|---|
| 3 A.M. | Lowest blood pressure. |
| 4 A.M. | Asthma attacks most severe. |
| 6 A.M. | Hay fever and cold symptoms greatest. |
| | Rheumatoid arthritis symptoms greatest. |
| 7 A.M. | Rise in blood pressure greatest of day. Angina pains, heart attack, and strokes most likely. |
| | Sexual desire highest in males. |
| 9 A.M. | Peak urinary volume. |
| 3 P.M. | Peak mental performance. |
| 4 P.M. | Peak lung function (defined as largest amount of air exchange per minute). |
| 3–6 P.M. | Osteoarthritis symptoms greatest. |
| | If healthy, this is the best time to exercise. |
| 9 P.M. | Blood pressure begins to decline. |
| 11 P.M. | Allergic responses begin to increase. |

In the last 20 years pharmacologists and doctors, once they began to look for it, have discovered that certain drugs are better taken at one time of day rather than at another. This special field of investigation has come to be called chronotherapy. To give a couple of examples: Aspirin taken at 7 A.M. remains circulating in the blood for 22 hours, but when taken at 7 P.M. circulates for only 17 hours. An antihistamine called Periactine taken at 7 A.M. remains effective for 16 hours, while the same amount taken at 7 P.M. lasts for only 7 hours.

Scheduling can also be important because some drugs, like those used in chemotherapy, can be more harmful when given at some particular portion of a day. Here are a few examples.

*A Cure for Cancer*

Leukemia is a form of cancer in which certain white blood cells are overproduced. They become so numerous that they take up residence in abnormal sites in the body (the emigration is called metastasis) where their presence is so disruptive that the individual is eventually killed. The malignancy is initiated and perpetuated as a result of an abnormally great rate of cell division, so this form of metastatic cancer is treated by administering inhibitors of cell division. Cytosine arabinoside (ara-C) is a good cell-division curtailer and is used to treat leukemia. In animal studies, when mice are inoculated with leukemia cells they eventually metastasize throughout the animals' bodies and kill them. If the inoculum is not too large, treating the mice with ara-C will prevent this. But the size of the dose is critical: ara-C stops cell division in normal cells also, so if the dose injected is too large, the treatment turns from a lifesaver into a life terminator.

Medical researchers cognizant of the importance of the involvement of biological rhythms in almost all processes, began a study on the changing tolerance to ara-C by injecting overdoses into mice at several different times of the day; they discovered that more mice were killed by nighttime injections than when the identical amounts of ara-C were given during the daylight hours. Clearly, the sensitivity of the mice to this cell-division inhibitor changes significantly as a function of clock hour. That learned, the following experimental procedure was designed and executed: For four consecutive days, leukemic mice were injected with 30 mg of ara-C every 3 hours. Thus each day the mice received a grand total of 240 mg of the inhibitor (this group is identified in the table below as the "Controls"). This dosage schedule rapidly killed a great number of the mice. Another group of mice (labelled "Experimentals" below) received the same grand total dose of 240 mg per day, but the amounts given during the nighttime hours, when the mice were most susceptible to ara-C, were reduced to as little as only 7.5 mg; while daytime doses, given when the mice's toleration of the medication was greatest, were increased to as much as 67.5 mg.

| | A.M. | | | | P.M. | | | | Total dose in 24 hours (mg/kg) |
|---|---|---|---|---|---|---|---|---|---|
| | 2 | 5 | 8 | 11 | 2 | 5 | 8 | 11 | |
| Dosage (mg/kg) | | | | | | | | | |
| Experimentals | 7.5 | 15.0 | 30.0 | 67.5 | 67.5 | 30.0 | 15.0 | 7.5 | 240 |
| Controls | 30.0 | 30.0 | 30.0 | 30.0 | 30.0 | 30.0 | 30.0 | 30.0 | 240 |

*Table 3.1. A table comparing the timing of different dose sizes of ara-C administered to controls and experimentals. The daily amount delivered to each group was identical.*

The experimentals lived twice as long as the control animals. These types of results began to appear in the early 1970s, but because they were often published in journals not routinely read by medical researchers and practitioners, and because the medical profession at large was also unaware of how widespread rhythms operated in humans and the rest of the living kingdom, the discoveries tended to go unnoticed.

Animal studies always precede experimentation using humans. This success with mice gave a go-ahead to use human subjects (mice are active at night, while humans are day active; thus in extrapolating mouse results to humans the daytime survival rates should be highest). In an 11-year study of Canadian children with acute leukemia, half were given chemotherapy in the morning and the others in the late afternoon or early evening. The cure-success rate was three times greater in the latter group.

Clearly, first determining any organism's rhythmic sensitivity to a medication, and then basing a treatment schedule on the rhythm's presence, can minimize the undesirable side effects of the treatment, including the extreme circumstance of patient death.

There are several hypotheses as to why a cytotoxic drug (for example, a cell killer) is quite deadly at one particular time of day and not at another. One possibility is that the clock controlling the cell division of cancer cells causes replication to peak at a different clock hour than the peak time at which healthy cells divide. Such time differences are known: Careful studies in humans have found that the greatest number of ovarian cancer cells divide from 9 to 10 P.M., while the rate of non-tumorous cell division peaks between 11 A.M. and noon. Thus, a chemotherapeutic strategy that dictates scheduling of the heaviest bombardment at night should be the most effective.

Two clinical drugs, doxorubicin and cisplatin, are often used in the treatment of ovarian cancer. Using the two together, rather than separately, prolongs survival time that ranges from as short as 10 months to a maximum of about 36 months. Medical researchers wondered if the time the drugs were administered might somehow explain the wide survival range. They began with animal studies. Six different groups of rats were given identical toxic doses of cisplatin, each at a different time of day. Those receiving the drug in the middle of the night survived the treatment significantly better than those injected at the beginning of the night. Another six groups of rats were given doxorubicin on the same schedule and were found to survive the experience best in the middle of the day and worst in the middle of the night. Clearly, the time each substance is dispensed plays a very important role.

Timed administrations were then tried on women with ovarian cancer. One group was treated in the usual way, they were just given the drugs without a strict time-of-day regimen. Another group received doxoru-bicin at 6 A.M. and cisplatin at 6 P.M. The third group got cisplatin at 6 A.M. and doxorubicin at 6 P.M. The results were quite striking. No member of the first group survived as long as three years. Eleven percent of those who received cisplatin in the morning and doxorubicin in the evening were still alive five years later. And most successful was the regimen beginning with doxorubicin in the morning: 44 percent of the test subjects were still alive at the end of five years. While it is sad that any of the subjects died, that is the nature of the disease. However, the take-home message here is that medical chronobiologists made a major contribution to prolonging the lives of the ill.

*Stifling Asthma*

The internal diameters of the air passageways of healthy people undergo a daily rhythm in size: At nighttime the caliber is slightly (about 8 percent on average) smaller. The same rhythm is present in people suffering from asthma but it is much more pronounced, the constriction being signifi-cantly greater. In people with mild asthma the air flow to the lungs may be reduced as much as 25 percent between midnight and 8 A.M., and in se-vere cases the reduction may be as high as 60 percent (something near a 50 percent reduction is the average for asthma sufferers as a group); the time of peak constriction is 4 A.M. Sixty-eight percent of asthma deaths occur between midnight and 8 A.M.

Asthma is a chronic disease characterized by coughing, wheezing, and breathlessness. Seventeen million Americans suffer from the disease; women account for 55 percent of the cases. Histamine is a small molecule that causes the constriction of the tubes in which air is circulated throughout the lungs and there is a daily rhythm in its liberation into the blood. The peak of this histamine rhythm coincides with the peak in breathlessness: 4 A.M. This can be demonstrated experimentally: When asthmatic patients were challenged with a histamine aerosol at different times of the day and night it was found that the airways reacted maxi-mally at 4 A.M. There is also a daily rhythm in the amount of epinephrine circulating in the blood. This substance helps keep the airways open and is at its lowest ebb around 4 A.M.

The standard treatments for asthma involve the use of bronchiodilator and anti-inflammatory medication, often administrated as an inhaled aerosol after an attack has begun. While these are very effective, their ben-efit is short-lived. Even when administered just before retiring at night, they will not retain their usefulness until the dangerous time in the wee hours.

Seventy-five percent of asthmatics awaken at least once weekly because of breathing distress, so the medical profession, now armed with its knowledge of the rhythmic nature of disease, is developing and testing timed-release delivery systems that will stem attacks in the early morning. Taking a controlled-release oral theophylline at dinner time shows great promise.

*Arthritis*

There are two kinds of arthritis, both of which are painful and crippling. Rheumatoid arthritis is a disease of our immune system. This system, when it is functioning normally, protects us from harmful invading organisms and other maladies. But when it malfunctions it mistakenly attacks, and may totally destroy, the joints between our bones. Our white blood cells are the functional unit of the immune system and they circulate in the greatest numbers in the early morning. Thus, the pain of this form of arthritis is greatest when we awaken, making getting out of a comfortable bed doubly difficult.

Osteoarthritis is caused by the destruction—usually by some physical activity you have undertaken in the past—of cartilage in bone joints. It is especially common in football and soccer players where cleat-bearing shoes firmly plant a player's foot in place and hold it there when he is then hit from the side; knees aren't made to bend to the right and left. In my football days I remember my mother lecturing me, "you'll be sorry later in life," and my father countering with, "give him a chance to be Saturday's hero." I never was Saturday's hero, but I did get a metal medal for my efforts: a titanium left hip joint! As a result of my brush with osteoarthritis I can say with great authority that osteoarthritis pain is usually most severe in the late afternoon and evening after a day's activity has further stressed sore joints.

Non-steroidal anti-inflammatory drugs (the unpronounceable NSAIDs)—there are many—are used to treat both kinds of arthritis. When used for rheumatoid arthritis, it is usually wise to take a timed release NSAID in the evening so that a good supply will still be circulating to quell the morning pain. People suffering from osteoarthritis should take their pills four to six hours before the worst pain of the day (afternoon and evening) would otherwise begin. This account would not be complete without mentioning that NSAIDs can sometimes play havoc with one's digestive system and liver. Be careful: In 1997 NSAIDs were claimed to be responsible for the deaths of 16,500 patients. Pain is very subjective and thus hard for an investigator to measure, but one study has shown that taking NSAIDs at 8 P.M. reduces the side effects, while the opposite is true for ingestion at 8 A.M.

## Hypertension

There is a relatively low-amplitude clock-controlled rhythm in blood pressure that was discovered by keeping people fasting and in bed all day. The rhythm was found to persist even in a comatose subject. As would be expected it is embedded in the more obvious overt blood-pressure rhythm created by being awake and active during the day alternating with lying asleep in bed at night. Both clock and behavior create low blood pressure at night and much higher pressure during daytime. Awaking and rising from bed in the morning creates the steepest climb in blood pressure each day. This stress on the circulatory system of older people too often results in strokes and heart attacks (thus, early birds, while you may feel like hopping out of bed, go slow at first or you might not get to your breakfast worm). That is one of the reasons that many people die in the early morning. It is also a reason for taking antihypertensive medication.

Such medicines are usually taken on awakening, but because they may take an hour or two to reach full effectiveness their action can miss the most dangerous time of the day. Again, the ideal situation would be to take a time-delayed release pill before retiring that would then go to work a couple of hours before one awakes when its benefits are greatest. In 1996 the FDA approved the chrontherapeutic drug Covera-HS. It is a package of the well-tested high-blood-pressure reducer verapamil hydrochloride, compounded in such a way that the active material is not released for several hours after being swallowed at bedtime. Thus, the medication will be at its most potent concentration at the most auspicious time.

There are many more similar examples that could be mentioned here but I am sure you have gotten the gist of the story. In a few year's time bioengineers undoubtedly will have produced chips that could be implanted in humans and that would control the release of medications at their most propitious times.

## Testing for Diabetes

A blood test is used to identify individuals with Type 2 diabetes (a disease characterized by an insufficiency of insulin, a hormone that promotes the absorption of glucose into cells). In usual testing, an individual fasts overnight, skips breakfast, and has blood is drawn in the morning. A glucose value higher than 125 mg/deciliter is indicative of the disease.

But over the years the number of people needing to be screened increased greatly. Thus, it became necessary to measure people in the afternoon also. As well this group was required to fast, the shortest interval being at least four hours before blood was drawn. Eventually, it was noticed that these people as a group had fewer diabetic members than those tested in the morning. Retesting afternoon people in the morning revealed that many actually suffered from diabetes. The living clock is the culprit here; the story goes like this.

There is a daily rhythm in the amount of glucose circulating in one's blood; it is highest in the morning and lowest in the afternoon. The 125 mg/dl cutoff level was set when the measurements were typically made in the morning. But thanks to the daily rhythm, the blood glucose levels of many actual diabetics never reaches 125 mg/dl in the afternoon, as was discovered when they were remeasured in the morning.

The solution to this problem is easy: Just set a lower afternoon glucose cutoff point to take into account the daily rhythm.

While the following story is not germane to this chapter, it should be edifying to everyone. I once heard Lewis Thomas (a member of the National Academy of Science and a former president of the Sloan-Kettering Cancer Institute) give a lecture on the success rate of patient care. He stated that the body heals itself without the helping hand of the medical profession in 90 percent of all illnesses. Beyond that doctors can help 7 percent of the time but can do nothing with the remaining 3 percent. However, these noble humanitarians allow the patient to give them full credit for a 97 percent success rate—a bedside manner that helps counter patients' shock on seeing the size of their bills.

# FOUR

# Jet Lag can be a Drag

## General and Specific Problems

W e are creatures of habit. Most people's day is organized around the eating of three meals, with each meal pretty much scheduled to the same time of day. We also tend to retire and arise on the same schedule.

On top of these basic biological dictates is the imposition of earning a living, an endeavor usually accomplished from 9 A.M. to 5 P.M. There are several niggling reinforcements associated with this schedule. Steady employment is difficult if one habitually comes late to work or leaves early; and long lunches are out unless you happen to be eating with the boss. I guess it is safe to say that individuals and society have forced themselves and itself into a rut . . . but this is not necessarily bad. Following a routine makes life a bit easier and much more efficient: It is not necessary to make decisions over the timing of fundamental events like eating and sleeping, and much more can be accomplished when we work in focused groups. And notice how a routine benefits older people, especially those whose minds are not functioning as well as they used to; they stick pretty much to rote behavior and thus glide through each day with only a minimal need for decision making and leadership. Of course, the rigidity can be overdone: Try and find a restaurant that will serve you a bowl of breakfast cereal at 7 P.M.

It may seem that our quotidian slog is self-imposed, but that is only partially true: Our living clock plays more than a minor role. Have you ever tried to force yourself to sleep at 11 A.M.? If you've had a good night's rest you can't do it. You can force yourself to stay awake, but only for a few days at most, then you begin to hallucinate and drop off to sleep no matter how hard you try to avoid it. The same is relatively true for eating. Only Dagwood Bumstead can get up in the middle of the night to eat one of his enormous sandwiches and still not have indigestion for the remainder of the comic strip. You too can binge in the wee hours, but you won't feel well afterward because your living clock has not prepared your stomach to receive food at that time. Cramps and discomfort can be the reward for breaking your rut this way. Remember that beneath your overt behavior is your clock that functions very well in preparing you and your body parts in advance for the next periodic event facing you, such as regular mealtimes.

Let me digress for a short paragraph to drive home the point about a main function of the clock: preparing a person for coming periodic changes in the environment. Just for fun I'll take up a problem Count Dracula has. He sleeps in his coffin during daylight hours and ventures out after darkness falls for a blood meal; legend has it that if he ever sees the sun he is kaput. How, from his dark crypt, would he ever know when the sun had set so he can safely venture out? It would be life threatening for him to even dare to peek out a window. Fortunately for the count, his living clock tells him when it is safe to venture forth. Another example of the same comes from the traditional relative of Dracula, the bat. Many live upside down so deep in caves that sunlight never reaches them. They too use their clocks to tell them when night has fallen and it is time to leave the cave to forage.

Here is a real-life example of the conflict that results when you try to override your clock's dictate, a lesson first learned by flier Wiley Post during his historic circumnavigation of the globe in 1931. Let's say that one summer you decide to visit an old friend in Norway so you begin by buying an airline ticket from Boston to Oslo. The ticket times indicate that you leave Boston at 5 P.M. and arrive in Oslo at 7 A.M.; you think to yourself, a nice overnight flight. But Oslo time is six hours ahead of Massachusetts and your living clock is set to follow the time in Boston where you live. You are about to have the travel problems that I will describe below. To help you follow my story, use the following comparisons of time between Boston and Oslo.

```
Boston ↓6 . . . . . M↓ . . . . 6↓ . . . . N↓ . . . . 6 . . . ↓ . M↓
Oslo   . M . . . . . 6 . . . . . N . . . . 6 . . . . M . . . . . 6.
```

In this comparison, M is short for midnight; N for noon; 6's are for 6 A.M.'s and 6 P.M.'s; the falling arrows drop at the hours I discuss below; and each dot represents an hour of the day that is not represented by a letter, number, or ↓. You take off at 5 P.M. (the first falling arrow above) and are fed a delicious dinner (because you are flying SAS) at 7 P.M. You watch the movie and then fall asleep. The next thing you are aware of is the cabin staff waking you up for breakfast. You check your watch and see that it reads midnight. Egad, it's midnight and I'm about to be served breakfast! "Yes sir," says the flight attendant, "It's just about breakfast time in Oslo and we'll be landing there in one hour. You look out your window and find that the sun is up. This is because you are flying against the rotation of the earth and because you have travelled from 40.4° to 60° north latitude where the summer days are much longer. You think to yourself that you probably overpaid something like $20 for that breakfast, maybe you should eat it even though you are not hungry and are still a bit tipsy from the free wine and drinks you had a short while ago. My advice is don't eat it: Your stomach clock is now signaling midnight so it is not prepared to receive food. But nobody ever listens to me, so you eat. You land at 7 A.M. Oslo time (next falling arrow that is at 1 A.M. your time) and deplane, overly full, with only two hours of sleep, and your clock saying "you fool, you should be asleep in bed." But, of course, you can't be, you're standing on hot tarmac. Soon your friend picks you up, drives you around that beautiful city so you can get your bearings, and then takes you to her home. It is fortunate that you do not have to drive because your clock is about to signal the time of day when your efficiency and judgment rhythms are at their lowest point . . . in fact, it's the most frequent time that people die! While you are freshening up, she prepares a typical lunch: one of those fabulous Scandinavian meals built on a foundation of butter, mayo, cheese, egg yolks, cream, pastries, sausages, and so on—a meal high in saturated fatty acids and cholesterol . . . a real artery clogger. Should you eat the meal? You'd be crazy not to: This kind of food is spectacular! If you listened to me earlier and skipped the breakfast served on the plane, this meal would be your reward. Also, this isn't a bad time to eat: Boston time is 7 A.M. (↓) so your body is readying itself for breakfast. Damn the arterial plaque; dig in.

And the same is true at dinner time (7 P.M. Oslo time is 1 P.M. Boston time[↓]): Your body is again ready for a meal, albeit lunch. At 11 P.M. Oslo time your hostess has had a full day so she retires for the night. But it is only 5 P.M. your time and even though you have only slept for 2 of the last 24 hours you now have your second wind. Pity. Should you go out on the town for a while and maybe even have a snack? (Your clock is signaling that it's near dinner time back in Boston.) Except for the night-owl Vikings, Oslo is going to bed; maybe you should think about some shut-eye also. You try, but you can't force yourself to sleep. Let's say you toss and turn until 10 P.M. your time (↓) and then fall asleep. At last, rest, wonderful sleep, but in three hours your hostess wakes you up; it's 7 A.M. (last ↓) in Oslo: time to eat, get out, and see more sights. Bloodshot-eyed and confused, you arise and ask something silly like, "What year is it?" The symptoms of jet lag have begun.

Now let's see what would have happened if we had instead flown westward, say to Honolulu, where the time is five hours later than at Boston.

```
Boston   6↓ . . . . N . . . ↓ . 6 . . . . . M . . . ↓ . 6 . ↓ . . . N . . ↓ . . 6
Honolulu . . . . 6 . . . . . N . . . . . 6 . . . . . M . . . . . 6 . . . . . N .
```

Our plane leaves at 7 A.M. (↓), we are fed lunch at noon, and arrive in Honolulu about 4 P.M. (↓) our time. By the time we find our bags it will be dinner time at home for us and lunchtime for Hawaiians: good living-clock time for both of us because their and our stomachs recognize this as an occasion to receive food. But then problems in a lack of synchrony begin. By 11 P.M. our clock's time, we are ready for bed, but a 7 P.M. luau has been planned for our group so we decide to pull a late nighter and get there on time (but it is midnight our time when it starts). It is great fun but we must have eaten three-quarters of a roast pig, and I can't even guess how many Mai Tais we consumed. I'm not sure but I think we got to bed at 11 P.M. Honolulu time, which is 4 A.M. Boston time (↓). Sleep at last. But four hours later (↓) we get a wake-up call from our internal clocks and our colons. Having gorged our bodies with food and drink at an inappropriate time we have a level of stomach distress that laughs at Tums therapy. On top of all this it's still pitch dark outside so there is no sense in trying to recuperate on the beach. Eventually the sun does come up and Honolulu comes to life; about 10 A.M. Honolulu time (3 P.M. ours; last ↓) we stagger out onto the beach, spread our blankets, fall asleep, and manage second-degree sunburns!

I have just run through the problems that arise on the first day of east or west rapid travel. Depending on the individual, and more importantly on how many time zones were crossed, mild physical and mental problems arise that can last for several days. A variety of maladies can be experienced: weakness in the legs, excessive tearing, either diarrhea or constipation, sleep disruption, an impairment in decision making, and a general feeling of malaise, just to mention a few. Whatever particular ills you experience, it is called jet lag, and somewhere between two-thirds and three-quarters of people suffer from it. The symptoms last until your clock has reset itself to the new time zone. Obviously, the faster this happens, the sooner you'll be totally happy in your new location. At the end of this section I'll make some suggestions that will hopefully help speed up the process.

Before we get too deeply into this I will emphasize that jet-lag problems arise only after rapid (it doesn't happen taking the slow boat to China) east or west travel across time zones. Travelling north or south does not create jet-lag problems if one stays in the same time zone. Still, north-south travel can require adjustment for sunrise and sunset differences. Suppose you were living in Santiago, Chile in July during the short days of winter there. If you arose each morning with the sun you would be getting up somewhere around 9 A.M. But if you flew directly north (remaining in same time zone) to Boston and awoke at your usual time of 9 A.M., you would have missed sunrise by about four hours.

Many, maybe even most, people find it harder to adjust to a new time zone after travel in an eastward direction. Several explanations for this have been elaborated, but I will describe only the most obvious one: When you travel to the east, as we did in the Oslo example, night-time comes sooner than you are used to, meaning you stay up alone when the rest of the people around you go to bed, or you retire at the same time they do but cannot fall asleep. As mentioned before, you can force yourself to stay awake, but forcing yourself to fall asleep when not tired is not easy. That was not a problem in the Honolulu example, there the difficulty was simply staying awake for the luau. Staying awake past one's usual bedtime is what many of us enjoy on a Saturday night at home.

The "east-is-best/no-it-isn't" direction argument has been carried to many extremes. A recent study was made of the relative winning success of Major League Baseball teams after cross-country travel to away games. Statistical comparisons found that home teams scored 1.24 more runs than usual against visitors who had just flown eastward across three time zones for the match; presumably the reduced showing of the visitors was a result of their jet lag. East-coast teams traveling west do not seem to suffer as much from the trip. What has been the baseball leagues's reaction to this startling finding? They were underwhelmed! But if follow-up studies confirm this observation, and remembering that winning is everything, will west-coast teams begin flying west around the earth to get to the east coast in order to improve their hitting? Let's hope not.

Now let's look at the importance of jet-lag problems. In the Oslo and Honolulu examples we were just vacationing so our jet-lag difficulties were mainly only a nuisance. But if travel is associated with important business, the outcome may depend on the timing of the trip. I tell my students the following story. An Italian businessman must attempt to close an important business deal with a company in New York City. He leaves Rome's Da Vinci airport at 9 A.M. on a flight that carries him 4,200 miles across the Atlantic Ocean in eight hours and arrives at Kennedy Airport at

11 A.M. New York time, which is 5 P.M. according to his biological clock; follow his adventures below.

His luggage is lost but that doesn't matter because he has his presentation in the notebook computer under his arm. He meets his counter-

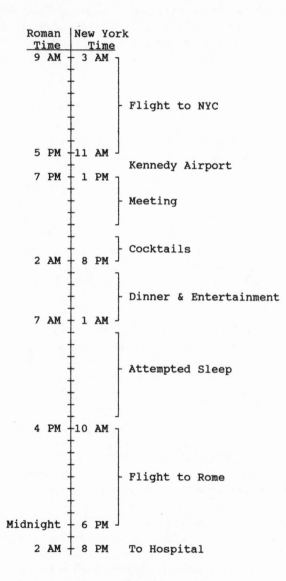

*Figure 4.1*

parts at 1 P.M. and they begin negotiations. The Italian business man's temperature rhythm is at its daily peak, and he is feeling especially fit. At this moment he has a double advantage because his business adversaries are experiencing their postprandial dip, the letdown many of us experience after midday. The Americans are putty in his hands, and he pulls off a great business coup just as his computer battery runs down. Although winning is everything, the Americans were good sports and take the visitor out for drinks before dinner. The drinks seem only to loosen the Americans' tongues (remember, alcohol is cleared from the body most rapidly at this time of day), and jokes begin to flow. However the Italian finds the experience quite intoxicating because it is after midnight his time, and the alcohol remains circulating in his blood, marinating his brain. After dinner and entertainment the Americans take their exhausted, now good buddy, to his hotel. He gets to his room, finds that the airline has delivered someone else's luggage, and falls into bed. Although he has been without sleep for 24 hours and feels wrung out, he has difficulty falling asleep because his living clock is signaling 8 A.M. Rome time. He tosses and turns and finally gives up and gets to the airport early to catch a 10 A.M. flight back home. At 4 P.M. the flight attendants initiate the cocktail hour, and our protagonist heroically downs a sufficient number of drinks to perpetuate the reputation of his countrymen. It is after 2 A.M. Rome time when he finally staggers into his house. He is intoxicated, exhausted, and in a state of confusion, but is still able to triumphantly blurt out to his wife a somewhat jumbled account of his business coup. Her response, "Quick dear, get me to the hospital, I'm going into labor" (here again, the living clock is calling the shots)!

That story had a happy ending for Italy, but the deal could have gone the other way if travel times and distances had been different. A true story drives this point home. John Foster Dulles, a former U.S. Secretary of State, deep into the usual frenetic shuttle diplomacy characteristic of that office since the beginning of the jet age, bungled negotiations with Egypt over the construction of the Aswan Dam, and the United States lost out to the USSR. It was a significant *faux pax* because the Soviet influence in Egypt that began then lasted for the next ten years. In reflecting on his lack of success, Dulles says that he had travelled back and forth across many time zones in the days and weeks before his arrival in Egypt and was thoroughly discombobulated due to jet lag when he attempted to negotiate. (Just what the country didn't need, yet another excuse for politicians to use when they fail to do their jobs properly.)

It should be clear that diplomats must take into account, and be wary of, jet lag in international dealings, otherwise regrettable consequences may ensue. The same is true for athletes, racehorses, and businessmen (not necessarily in that order).

The problem here is obvious, one's clock is set to, and by, the day/night conditions of your home time zone. When you move rapidly to a new zone, the clock must be reset, and the main resetting stimulus is, of course, the new ambient light/dark cycle as recorded by your eyes. Logically then, when we fly rapidly to a new time zone it is imperative that we immediately expose ourselves to as much daylight there as possible during the first few days. Remember when we flew to Honolulu, we arrived at noon local time, but our clocks were expecting only a few more hours of light before sunset. It might be fun to start vacationing immediately, step into a dark bar, and quaff down Mai Tais all afternoon, but don't do it; instead put on number 15 sunscreen, leave the sunglasses off (but don't try to count the spots on the sun), and stay outside. Then when the sun sets, it would be equally wise to stay out of bright artificial lighting and get to bed early. You probably will not sleep well, but tough it out in the dark rather than turning on the light to read. Your clock resets rather rapidly, but it drags your rhythms into the new phase rather slowly, depending on the kind of rhythm it's tugging on (the sleep-wake rhythm resets rather rapidly; some excretory rhythms take a very long time), and how many hours a rhythm must be rephased (for example, how many time zones you traversed). Then, when the sun rises, get up and go outside—you get the idea.

On first arrival, when, and if, mealtimes in the new location happen to correspond to any of your mealtimes back home, even an old breakfast time with the new dinner time, it is safe to indulge yourself a bit. If such correspondence does not exist, eat lightly, and avoid or skimp on the wine. Doing this will help you avoid stomach upsets. But after about a day and a half your rhythms will have begun to rephase so forget the previous meal-correspondence attempts since they will no longer exist, but it is still wise not to binge, and resist over-indulging on spirits, even though you may be strongly tempted to try new exotic drinks. And watch those deadly rum-based punches, they may taste like your morning glass of fruit juice, but treat the punch part of their name as pugilistic.

The alternative to this sensible approach is to damn the torpedoes and start your vacation full tilt the moment you step off the plane; the rewards for this mindlessness can be Montezuma's revenge (his vengeance reaches far beyond Mexico) or constipation, cramps, horrible hangovers, automobile accidents, inability to concentrate, et cetera. You may also make some really stupid mistakes, like former President George Bush did in 1992: After country hopping for several days, when he arrived in Tokyo, macho man that he is, in spite of his jet lag, he foolishly played tennis that afternoon and then went to a state dinner where he threw up on the prime minister's foot. After which, still chest thumping, he actually said to the Prime Minister, "Why don't you just roll me under the table and I'll sleep it off." As you see, even presidents can come unglued when they try to beat their living clock.

Before getting into the scientific remedies for jet lag, I'll start with some home countermeasures. If your trip is short and important (like the Italian business man's overnight excursion) try to schedule it so you will be in peak form for the important event. But being a prisoner of airline schedules, that can be difficult. If the trip is extremely important, you could arrive a sufficient number of days early so that your clock has an opportunity to readjust to that time zone; but that requires a rare, wise, and informed boss . . . otherwise those extra days are considered vacation time—and your company loyalty may even be suspect. You might try to preset your clock at home and start living your destination's day/night schedule a few days in advance of the trip, but this attempt at preadaptation can result in demerits when it means you must get to work late or leave early. Also, needless to say, if you abandon a civilized quotidian and stay away from home at weird hours, your family life will be disrupted. On top of all this, if you make the change to the new time-zone hours while still at home you get to suffer all the symptoms of jet lag in the presence of employer and family. It's tough and risky, probably not worth a try. I attempted it once as an experiment and almost got fired and divorced, but my junkyard dog still loved me (I'm the one who fed him).

The other side of this coin is to attempt to keep your home schedule in the new time zone. Former President Lyndon Johnson used this trick as often as possible and was sometimes successful, but that feat is easier to do when you are the president of a nation with an enormous fighting machine and enough nuclear weapons to destroy most anything several times over. He also had his own plane (Air Force One) to buzz around in and no need to sit around in airports for hours at a time. For non-presidents there are small enclaves here and there that are sympathetic to people who would like to try that trick. For instance, in New York City there is a hotel that has six "circadian rooms" that are equipped with dialable light/dark schedules for people visiting the city for just a short time who want to remain on their home time-zone schedule.

Now to what science has to offer the new traveller. The most obvious, the most foolproof suggestion, and the one guaranteed to work is: Stay home. But for the others of you who decide to travel, here is how your living clock will respond to the new day/night schedule at your destination. In the following sketches, all the bars are 24 hours long and represent day/night alternation. The stippling stands for a total of 16 hours of day-

light, separated by 8 hours of night (the solid black). For the moment ignore the smiley faces and the brackets. Let's say the upper bar, labelled
"you," is the day/night cycle you presently are living in and to which your
living clock is adjusted.

You plan a two-week vacation that will require you to fly westward
across four time zones. The lower bar, labelled "westward," represents the
timing of the day/night cycle at your destination relative to the one in
which you are now living. When you arrive at your new venue, your watch
and your living clock will be four hours out of synchrony with the time
there; your watch will say 7 P.M., but it will be only 3 P.M. there. Nightfall
there will come four hours later for you, as emphasized by the bracket.
Your exposure to these new hours of light will cause a phase shift in your
sleep/wake rhythm to the right so as to synchronize it with the new vacation venue. In order for you to adjust to the new time zone, your living
clock must be delayed four hours, and it has a special mechanism to do
this. For the sake of simplification I have represented that phase-setting
mechanism by a string of smiley faces.

The faces come in two versions, dark and white. When at least one
white resetter-face is illuminated (in this case seven are), it causes your activity rhythm to rephase to the right. This change takes place at a rate of a
little more than an hour a day and stops when sunlight no longer illumi-
nates any white faces. At that point your sleep/wake rhythm has adopted
the new western day/night schedule.

Now let's say you have stayed out west long enough for your living
clock to totally adjust to the time zone there, and now you are about to return to your eastern home.

Here shows the alignment of timing between you and the eastern
day/night cycle when you first arrive.

When your wristwatch says 4 P.M., the time there is already 8 P.M. Home
again, the last four hours of what had been darkness for you out west

(here bracketed) are now illuminated, as are the black smiley faces of your clock's phase-setting mechanism. When light falls on at least one of them they push your rhythms to the left in the diagram. When all of these heads are no longer illuminated, your sleep/wake rhythm is now in sync with the eastern light/dark cycle.

Although smiley faces have been used to alleviate the reader's need to suffer an excursion into biological terminology, and although I oversimplified the story by leaving out some details and using an ideal situation, my description is otherwise accurate for humans and other species on which detailed studies have been made. What I have depicted is the main way all organisms can adjust their daily rhythms to new time zones.

To this idealized schema of the mechanism I'll now add a complicating factor. The figure below focuses on the spot where the points of the double arrows separating light and dark heads actually touch; this is a critical locus where an organism's clock-setting instructions change between delays and advances. With this fact in mind, use the diagram below to see that when we jet across a great number of time zones, both light and dark faces are simultaneously illuminated (at A and B).

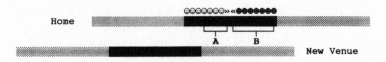

This creates a kind of Dr. Doolittle push-me/pull-me situation in that smiley-face forces try to set the clock forward and backward at the same time—it is reset to the right by light hitting the previous dark segment bracketed by A, and set to the left by the light falling on the area bracketed by B. The final resetting direction each day is determined by the relative numbers of light and dark smileys being illuminated; and because the B segment has more than the A portion, the rhythm eventually adjusts to the left. But, as long as even a few white faces are illuminated they shove the rhythm a bit to the right, and this wrong way also has to be countered by the dark faces. With time, however, there is eventually synchronization with the new day/night cycle, but it takes longer when the smileys are involved in a shoving match.

So, what is recommended for us when we put ourselves into a situation like the one shown immediately above where both kinds of smiley faces are simultaneously illuminated? One option is to accept the fact that adjustment to the new time zone will take quite a while; another is, while a bit fanatical, to speed up the adjustment by staying in the dark or wearing a sleep mask during the interval that the white faces would be exposed to light (at A above); and a final option, which is probably the least effective, would be to at least wear dark glasses during this interval. A confounding problem with the last two options is that the exact change-over point between phase advance and phase delay (where the double arrows touch in the drawing) is not known with any certainty for humans. Then when do you take off the sleep mask or dark glasses? The hour usually suggested, but there is still little evidence establishing the accuracy of that point. Additionally, it may be that each of us has a somewhat different change-over point. We must wait until sufficient experimentation gives us better information.

It is staggering to think of all the places we can visit on earth, and thus the many calculations each of us would have to make to figure out when to put on dark glasses or when to take them off; when to stay inside or go out into the sunlight, et cetera. Happily, this work has been done for us: An inexpensive, paperback book (just slightly too large to fit in a pocket) is now available. It is called, *How to Beat Jet Lag.*

Sleep masks are just one kind of crutch. Headlights are another. I am not referring to the lights on your automobile that illuminate the road in front of you. These are $379 headlights, battery-powered affairs that sit on your head and shine light into your eyes. They are intended for use in dark situations created while flying when the light smileys need illumination to hustle up your adaptation to a new time zone. Needless to say, unless you are the extrovert kind who likes to wear lamp shades on your head at parties, the headlamp adornment can be a source of embarrassment until the country finally accepts it as a science/fashion statement. To reduce ostracism while wearing one of these, try burying your head in a newspaper as if you were reading. Nevertheless you still might have to suffer comments like, "Please sir, turn out your headlight, others in the plane are trying to watch the movie." So what. There are outspoken Luddites everywhere who aren't open to new ideas.

### A Knee(d) to Know

Clearly, the above account outlines the critical importance of getting light into the eyes at the proper times to readjust one's body clock to a new day/night cycle. Believe it or not, there is a claim that shining light on the area behind the knee joint is just as effective. In demonstrating this ability the temperature rhythms of volunteers were studied in constant conditions for four days. Starting on the second day each subject received a three-hour exposure to a blue-green light, focused on the area behind the knee joint. Depending on just when at night it was applied, and for how long, the test light set the temperature rhythms to either an earlier or later time of day!

## Sham Clocks, Playing Tricks with Time

There is more to the readjustment-of-rhythms story than just exposure to light during new nighttime. An experiment providing some of this information was done in Spitsbergen, a Norwegian island in the Svalbard Archipelago. This island lies at about 78° north latitude—just 12° south of the north pole. In the summer the sun never sets there and the day and night temperatures remain about the same. The sun does circle around the horizon once each day (meaning an astute observer could use its position to tell time), but more often than not the sky is overcast so the sun is obscured. Thus, in the high latitudes, the environment is naturally, roughly constant.

Nineteen subjects were flown from Britain to Spitsbergen and divided into three separate camps. Camping equipment and everything else was provided for them, including watches. Unbeknownst to them, all the watches in one camp signaled the passage of 12 hours in 10½ real hours; and in another camp they indicated 12 hours in 13½ actual hours. These subjects were told to carry out their normal routines (using their bogus watches, of course). None realized that they were actually living 21, or 27-hour days and were thus, overtly, happy campers. They were quite aware, however, that they were part of a biological experiment, because they had to provide a urine sample at roughly two-hour intervals throughout most of the day (and less frequently during periods of sleep). A third group, also in a separate camp, used normal watches. This pee-in-the-bottle exercise carried on for about six real weeks. Then all three groups broke camp and returned to Britain.

The results were not spectacular or consistent. Of the seven subjects who lived on a 21-hour routine, the excretory rhythm of only one quickly adopted the spurious day length, but later two more were fooled by the sham watches. Two other subjects retained their normal 24-hour period, while the other two became arrhythmic.

On the 27-hour routine two of the subjects adjusted to it very quickly, one more adopted it after a couple of weeks, and two others adopted the 27-hour period water-volume rhythm, but not the potassium-excretion rhythm. The rhythms of the others in the group ignored the watch's disingenuousness and stuck to 24-hour periods.

The third group, even though living in quasi-constant conditions, had strict 24-hour rhythms, guided, of course, by their honest watches.

In all three camps, the temperature and sleep/wake rhythms of the volunteers followed the time given by the watches they wore. There were more subtle results obtained from this experiment, but for our purposes here, it is enough to see that artificial stimuli (such as, sham watches) may, for some people, also serve as time-setters of our living clocks.

Interpersonal relationships also serve, to some extent, to control the timing of our rhythms. This should not be too surprising since we are, after all, social animals. To focus mainly on social stimuli, four subjects were placed together, in constant conditions in the underground bunker described in Chapter 2. The periods of their sleep/wake rhythms lengthened a tad, but in essence the four stayed synchronized—except for one early bird who awoke a bit sooner than the others. During the first 16 days in the bunker they stayed pretty much tuned to each other in this pattern, but on day 17, even though the lighting conditions in the bunker did not change, they broke synchrony. The early riser adopted a period of 24.1 hours, while the other's rhythms stretched to 27.2 hours. On day 20 another subject began to shorten his period so that when the investigators overseeing this exercise broke into the bunker to announce that the experiment was over, they found all four at the table eating; however, one was having breakfast, two were having lunch, and one was eating dinner.

There are other confounding problems. Because it is the day/night cycle that plays the fundamental role in synchronizing our rhythms to a new time zone, laboratory simulations using light/dark cycles provide just as good, and a much less expensive substitute for jetting subjects around the world. In addition to saving costs, a second advantage is that the experiment is not contaminated by the fact that human subjects are flown to new interesting countries where they are inevitably distracted by social events, cathedral tours, novel foods, and exotic drinks, et cetera. Figure 4.1 shows how a typical experiment is done and presents an example of the kinds of results that can be obtained.

On the first line at the top of Figure 4.1 you see that this person arose an hour after light-on (= dawn) and retired at 11 P.M. (using a small reading lamp after the major lights were turned out). The peaks of the temperature rhythm and the minima of the sodium-excretion rhythm came at 8 P.M. and 9 P.M., respectively.

The rhythms of this subject were followed for three days to establish their relative phases and then the light in the room was turned on and off five hours earlier. As far as day and night are concerned, this change was tantamount to flying someone from Boston to London. On the first day of the change (labelled as one), note that the subject was awakened when the light came on at an unaccustomed time, and then nodded off and awakened a few more times. When the light went off at 2 P.M. the person

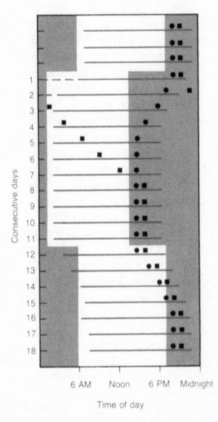

6 AM    Noon    6 PM    Midnight

Time of day

*Figure 4.5. A diagrammatic representation of subjecting a subject to a 5-hour phase shift in an easterly direction, and then after 11 days, returning him back home. The bars indicate intervals of being awake, the dots the temperature maximas for the day, and the squares the minimas of the daily sodium-excretion rhythm. The shaded areas are times of darkness.*

stayed awake until almost his normal bedtime; the other two rhythms did not yet begin to change. But by the second day all the rhythms had begun to rephase. What is important to notice here is that the sleep/wake and temperature rhythms began to move to an earlier time, while the sodium-excretion rhythm began to rephase by going the other way. The first rhythm to adjust to the London schedule was the sleep/wake one, the temperature rhythm was next, and the excretory rhythm last.

When one is totally in tune with his environment, all his rhythms are in the proper phase relationship with one another, as seen at the top of Figure 4.1. Until then his body is not functioning just right, and this can lead to some minor jet-lag symptoms, but none as severe as experienced during the first couple of days. Thus in the case above it wasn't until day eight that the dissonance of the three rhythms had been corrected and the subject was totally fit again.

Beginning on day 12, the light/dark schedule was set back to the starting regimen, and the rhythms began to adjust. As before, it took only a few days for the sleep/wake rhythm to rephase, but look at the difference in the other two. On the simulated flight to London it took the temperature rhythm five days to adjust, and the excretion rhythm eight days (and it did so by moving in the opposite direction to the temperature rhythm). But now after flying in the opposite direction, both rhythms rephased in tandem in the same direction, and in only five days. Thus, our stationary traveller adjusted more quickly going in this direction.

Another point that should be made quite clear is that the responses to transmeridional travel is an individual thing. Probably something like 20 to 30 percent of people aren't bothered much at all by it. Some people adjust rapidly after a trip, and some do not. Finally some go east better than west, and vise versa. All of the above is also characteristic of shift-work alterations, as will be described later in the chapter.

The travel time to a destination where jet lag can become a problem can add to the symptoms, especially if the flight is overnight, and especially for those in snug tourist class where sleep can be almost impossible. I worry most about the pilots; many who will have to fly soon again with only a minimal rest time. Remember Charles Lindbergh's diary entry near the end of his 33½-hour solo transatlantic flight, "My whole body argues dully that nothing—nothing life can attain—is quite so desirable as sleep." At that moment concerns like being the first to fly the Atlantic, and expectations like cheering crowds awaiting him, were the farthest thing from his mind. Lindy was lucky and made it. Well, in spite of laws, rules, and conscientiousness, pilots do sometimes fall asleep . . . after all they are human. A few years ago a handful of airlines finally began to let pilots nap officially (but that change did not become part of airline marketing strategy). British Airways and Air New Zealand were first to try this. Others followed. When the public did learn of it, airlines touted (when asked) up to 40-minute naps on the less demanding parts of long flights as a safety feature to ensure alertness during the most dangerous part of the flight, which is of course landing. That seems a good decision to me: I like smooth landings. The nap policy was accepted by government regulatory agencies and now berths are even built into the cockpits of larger planes for pilot snoozes. In fact, as I write this (April 1999), a Delta Airlines captain who took off for Tokyo landed his plane in Portland, Oregon rather than crossing the Pacific because the bed provided in his cockpit was uncomfortable!

*Clocks at War*

Knowledge of human rhythms was used by the Air Force Command in attacks in Serbia in 1999. For the first time in its short history Americans carried out sustained bombing attacks on the enemy from its own soil. American B-2 bombers, valued at $2.2 billion each, took off from Missouri, flew eastward for 15 hours, dropped their 2,000-pound bombs on Belgrade and vicinity, and returned to Missouri. Considering how mean I feel after just a 13-hour non-stop flight, this appears a good pilot-incentive strategy to me. The leadership's wisdom also included resetting each pilot's living clock before attempting one of these $441,270 flights. A week before each flight pilots begin to adjust their sleep patterns to maximize their alertness during the short bombing runs over their targets. Some pilots even added to their clock-driven "up" by having a cup of coffee 20 minutes before reaching their targets. Then, as reported in the newspapers, they flew back home, picked up their kids after school, and napped for a couple of hours before dinner.

*Winter Blues, Let There be Light*

A malady different than jet lag is also treated by shining artificial light into one's eyes. It is estimated that 1 in 10 people in Alaska, 1 in 100 people in Florida, and something like 10 million Americans suffer from Seasonal Affective Disorder, SAD for short. The season in the appellation is winter, when the number of hours of daylight are few. The disorder is depression, anxiety, irritability, decreased energy, a petered-out libido, and sadness, making the SAD acronym very appropriate.

The best prevention or cure is to leave winter here and go to South America, repeating summer down there. Another source of help is to sit in front of a bright light for 30 to 90 minutes each day. Not everyone who is treated this way responds; but those who believe that the treatment will actually work have a higher success rate. About 60 percent of the subjects who receive the treatment in the morning, versus 30 percent of those who were exposed at night, find at least some relief in their SAD symptoms after four weeks of daily treatment. The light intensity used was about 20 times brighter (10,000 lux) than the standard illumination in a typical office.

It is very hard to run an adequate control that would show whether the light was actually responsible for the improvement, or whether the change was just a person's subjective belief that treatment with light will work (a placebo effect).[5] The obvious control, staring at an unlit light, will not work because everyone can see if the light is not on. An ersatz control was contrived and presented, which demonstrated how easily we can fool ourselves. Instead of a bright light, subjects were placed in front of a negative-ion generator for 90 minutes a day for a month and told that having ions blown in their face would zap away their SAD. Each day the machine hissed importantly, but unknown to the subjects, no ions were shot out. Nevertheless, the sibilance satisfied 36 percent of this group: They reported themselves cured.

## Shift Work, Poor Man's Intercontinental Travel

Up to now we have been talking only about jet lag, which by its very name indicates that an individual's living clock cannot keep up with rapid transmeridional transport. One may get the impression that jet lag is only a problem of those who can afford the high costs of intercontinental flights—the rich and those living beyond their means. That is not the case: Long before planes were invented there was rotational shift work, which, in its most common form of practice, is tantamount to jetting across 6 to 12 time zones in the blink of the eye. It costs nothing (it is an employer's preogative) and avoids the fatigue associated with a long flight with little leg room, but it produces all the same symptoms of jet lag.

Some businesses, in order to be able to afford owning expensive equipment, must keep it running day and night. Companies producing chemicals, steel, or electricity must run 24 hours a day. Hospitals and transportation services need round-the-clock employee services. And in good economic times some industries must work 24 hours a day just to meet the high demand for their widgets. Then add to this a nonstop stock market, the unsleeping internet, and all-night shopping. Adopting rotating-shift work schedules has been the usual solution in most industries.

Work shifts may vary from 6 to 12 hours; 8 hours is the most common. Workers may rotate every few days, or every week or month. It is estimated that 20 percent of the active, industrial work force is involved in shift work, and I have seen a figure as high as 27 percent for the United States.

Shifting once a week by eight hours seems to be the most common practice. It is also the most pernicious: It is tantamount to traveling once around the world in three hops every three weeks. That schedule is probably the most unsatisfactory one that could have been adopted because just as a person's rhythms should have about adjusted to the new routine, the work time is changed by another eight hours. Obviously that puts considerable and continuous stress on one's body and mind. Immediately after an eight-hour change, two deleterious, temporary, consequences befall: Productivity ebbs and the accident rate increases. And when a new shift results in moving one into working during the early, dangerous, A.M. hours, the chance for accident is even greater: Spectacular examples of this are Union Carbide's spill in 1984 of methyl isocyanate (an ingredient of pesticides) in Bhopal that killed 3,800 people immediately and another 11,000 thereafter; the technical errors made during a test of an enhanced safety procedure that caused the Chernobyl reactor to explode; and the nuclear power plant accident at Three Mile Island that took place at 4 A.M. just after a new crew had rotated onto that shift. It is estimated that in the United States industrial deaths and injury cost the country $1.5 billion annually. Here is one from the column "Are our faces red." A company called Circadian Information mails out periodic tips to workers on the graveyard shift for improving their overnight alertness. At 3 A.M. one morning, company workers addressed and mailed 130,000 such letters— all went out with incomplete addresses. The mailing was redone.

In sympathy for one's living clock, and considering that accidents would be less likely to arise if workers were not shifted, why not stop the practice? That seems wise on paper, but permanency, say on the night shift, is often not really stable: On days off these people usually join their families and friends in daytime activities, which, of course, start their clocks to reset. Also, permanent night-shift workers sometimes use it as an opportunity to take a second job, sometimes literally working themselves to death.

In addition to such negatives as poor quality products and increased accident rates are greater absenteeism, common employee turnover, and increased illnesses. Exhausted workers do poor work and complain that off the job they lack energy for recreation and romance, so they grow irritable both at home and work. Marital breakups in families with children are threefold greater if the wife works at night, and sixfold greater when the husband does so (but there is not that risk for childless couples). "It's hard to have a life when your hours are always changing," carps a union bigwig in concern for union members. A company in Indiana polled its employees, and they voted for fixed 12-hour shifts with everyone assigned to either days or nights. The company reports reaping great rewards: Employee annual turnover dropped from 32 percent to 9 percent, productivity rose, the safety record of that branch of the company rose from worst to best, and health costs plunged—a wonderful example of how important it is to keep our living clocks from becoming overtaxed.

Some people just cannot do rotation shift work because it is so stressful. Just exactly how dangerous it may be for humans is unknown, but the results from animal experiments are scary. When blowflies were subjected to a six-hour phase shift of the ambient light/dark cycle once every week, their life span was reduced by 20 percent. When the light/dark cycle for mice was reversed once every week their lives were cut short by 6 percent. It's hard to know whether such dire consequences can be extrapolated to humans, but it is known that gastronomical maladies, including peptic ulcers, are two to three times more common in shift workers than in the general population. But I suppose if you were to inform the CEO of a major company about these experimental results, his answer might well be, "Aren't we wise not to employ blowflies and mice?"

Entire populations are exposed twice each year to a one-hour shift as we spring forward to Daylight-Saving Time in April and fall back to God's time in the fall. Other than the media's self-mandated hype at these moments, only a few over-enthusiastic science types have attempted to study the affect of such a small change on human behavior and physiology. One published study used traffic-accident statistics and compared the accident rates on the Mondays a week before and a week after the one-hour shift to the crash rate on the Monday of the change in time (the swap was always made at 2 A.M. on Sundays). In the spring, with the government's mandate that we all lose an hour's sleep, the middle Monday accident rates were increased by about 8 percent. In the fall, after an additional hour to luxuriate in bed, the rate fell by 10 percent. Both of these differences were statistically significant, meaning that they probably did not happen by chance. That significance, however, says nothing about the role of the living clock; the best guesstimate is that it was due to the loss of sleep in the spring (asleep-at-the-wheel syndrome) versus getting instead to "crash" on one's pillow an extra hour of sleep in the fall.

## MELATONIN, A WONDER DRUG

We live in a society interested in drugs, both the medical and recreational kinds. It is only natural that people ask if there isn't some chemical that we can take that will quickly change the setting of the hands on our living clocks so we can avoid the pain of jet lag and shift work. The answer is a definite maybe, and the perchance elixir presently popular is melatonin. Here is the background.

There is a cone-shaped, pea-sized area in the brains of higher organisms called the pineal organ. In amphibians its secretions cause them to change color; in some lizards it functions as a third eye; in humans—if we were to believe Descartes—it is the seat of our soul. Biologists, being less colorful and more disciplined than philosophers, recognize this area of the brain as a much less romantic entity: It is a gland that synthesizes and secretes the hormone melatonin.

Melatonin was discovered in 1959, and has since been found to be widely distributed throughout the living world, from algae all the way up to humans. It is claimed to have a variety of effects on humans, such as preventing or curing cancer, diabetes, cataracts, PMS, schizophrenia, epilepsy, Parkinson's, depression, Alzheimer's, and cardiovascular diseases. It may work as a male contraceptive in artificially high concentrations. There is even a claim that it may slow aging, yet another pretense too eagerly accepted by health-store shoppers.[6] The best counsel I can give here is to invoke the first law of hype detection: If it sounds too good to be true, it probably is. Melatonin, while it has its place in the pharmacopoeia, is often sold to the public as a New-Age snake oil.

But one property that has been proved beyond a shadow of a doubt is that melatonin has a mild hypnotic effect—it is a sleep inducer; it creates a very natural sleep including dream phases characterized by rapid-eye movements. That suggests that it should be useful to take if you end up in a new time zone where your living clock (still set at back-home time) interferes with your ability to fall asleep at local bedtimes, or when it awakens you before the rest of the locals normally arise. Additionally, the amount of melatonin circulating in our blood is rhythmic: It is most abundant at night. Light falling on our eyes tends to inhibit its release from the pineal, and darkness stimulates it; but the rhythm will persist in constant conditions—even in the eternal darkness that blind people must endure.

Beginning in the late 1970s it was found that the daily administration (always at the same hour) of melatonin would synchronize the circadian rhythms of birds, lizards, rats, and hamsters to a strict 24-hour period. Would it do the same with human rhythms, say, after transmeridional flights? Before investigating this possibility, let's discuss the safety of taking melatonin.

Investigators of the matter feel that melatonin is safe to take in spite of the fact that the Food and Drug Administration (FDA) has not approved its use on humans; that agency's statement is that those people who want to take it should do so without any assurance that it is safe or that it will have any beneficial effect. The normal level circulating in our blood at the peak of the melatonin rhythm is something around 100 picograms per milliliter of blood (a picogram is one trillionth of a gram). In the first experiment ever tried, 240 milligrams (that equals 240 million picograms—a stupendous overdose) were used; it made people very sleepy but also caused a terrible hangover the next day. The standard amount used in jet-lag reduction experiments is only a tad of that: It is as high as 5 milligrams and as low as 0.3 milligrams; all these concentrations usually work just fine as a sleeping pill, and thus far are not known to cause undesirable aftereffects. Taken orally, melatonin reaches a peak concentration in one's bloodstream somewhere between 30 and 60 minutes after ingestion.

Melatonin can be purchased in many health-food stores, but, because it is not commercially regulated, what is sold may be tainted with undesired impurities. Because it is a natural ingredient in some foods, it is considered a dietary supplement rather than a drug, and as such is not subject to the FDA's premarket approval as safe and useful. Also, the pills often contain only a small amount of melatonin, plus lots of other natural ingredients—"all known to be relaxing and soothing." Each pill in the bottle I'm holding in my hand at the moment, although labelled melatonin, contains only 15 percent melatonin. Other products contain melatonin extracted from cow pineals, and this has prompted some nascent Naders to worry that this could be a route of carrying mad cow disease to humans. I would warn that the bottom line here is that an insufficient amount of testing has been done at the time of this writing to insure the safety of melatonin. *Caveat emptor* is probably a worthwhile caution. Over-the-counter sales have been banned in Britain and France. In Canada it is available by prescription only.

So much for the disclaimers. Melatonin is taken routinely by many people and also used in jet-lag experiments. So, back to the question of whether it is worth taking to lessen the symptoms of jet lag. Many experiments have been undertaken, some with rather curious experimental protocols, and the results have been inconsistent. Last year saw the largest controlled study ever done on the subject. Two hundred fifty-six volunteers (203 men, 53 women), most of them physicians, evaluated the degree of their jet-lag symptoms after a flight from New York to Oslo, Norway. To quantify the severity, daily each subject used a scale ranging from 0 (feel fine) to 5 (quite bothered) to evaluate each of nine symptoms: fatigue, daytime sleepiness, impaired concentration, decreased alertness, recall problems, physical clumsiness, weakness, lethargy, and light-headedness.

After arriving in Oslo, for the next five days at bedtime 64 subjects took 5 mg of melatonin, 70 subjects took 0.5 mg, another 63 took 0.5 mg but they downed their's one hour earlier each day, and 60 took a placebo. None of the subjects knew what was in the daily capsule they took.

The outcome was rather unexpected. While all four groups suffered maximally on the first day in Oslo, and jet-lag symptoms declined toward zero over the next four days, there were no significant differences between the groups. It didn't matter how much melatonin one took, or the schedule used to take it, or whether it was taken at all! While the answer seems quite clear, the investigators of this study—always cautious—conclude that more work needs to be done.

Remember the blind subject described in chapter 2, who feared he would fail out of school because his sleep/wake rhythm was 24.8 hours long, and he therefore went through several consecutive days when he would fall asleep in class. Here is the good follow-up news: Recently he began taking melatonin slightly before a normal bedtime and it stabilized his rhythm to a 24-hour period. Here is more good news: He successfully completed his doctorate.

# FIVE

# Daily Rhythms in Single-cell Organisms

## The Rhythms of a Red-Tide Alga

We will now leave human-rhythm studies and return to other interesting members of the living kingdom; the next few chapters will build a background of knowledge leading to the discoveries of the clockworks described in Chapter 9.

The majority of chronobiologists work only on multicellular organisms, but, as you have already seen, single-cell plants and animals also undergo clock-controlled rhythms that are fascinating and fundamental to the arcana surrounding the living clock. We will look at a few more examples in this chapter.

Let's begin with algal rhythms that do not match the period of the tide; these rhythms are equal in length to the interval of a day.

One reads—usually annually—of the appearance of large colored stains forming in the ocean just offshore. If that is all that happened, even though it could be pictured by a TV camera, and even though the spot may be blood colored as it sometimes is, it would probably not be worthy of news hype. But it becomes network hoopla if onshore winds carry the odors of the stain to land and cause people there to choke and cough and complain of eye irritation. What has happened here is that environmental conditions had become extremely favorable for the growth of a particular species of alga, and it has enjoyed a reproductive frenzy that has culminated in a population explosion. Toxins and decay products from this enormous concentration of alga dissolve into the seawater and kill resident fish and invertebrates, while volatile substances diffuse into the air and waft onto the shore causing people to cough and sometimes even die. One of these algal-bloom organisms is *Gonyaulax* (gahn knee ALL axe), and when it blossoms it sullies the sea becoming a so-called red tide. A very fine scientist, the late Beatrice Sweeney, collected some of this alga off the coast of California, and, after learning what was needed to keep it alive and reproductively content in the laboratory, began to grow it in pure culture. She has shared her culture with many labs, and the alga has become a standard organism for experimentation in labs across the country and the world.

*Gonyaulax* is a single-cell, microscopic alga that can swim weakly by flailing around its two flagella (a flagellum is a hair-like filament like the tail of a human sperm). The alga is microscopic so one can only guessti-mate how many individuals it must take to cause several square miles of sea to stain red. In addition to reproductive prowess, it has other eye-catching qualities. I will compare (incorrectly) one of its characteristics to a human counterpart: It, like us, can change color when irritated. We blush; *Gonyaulax* flashes blue when whacked!

At night when a fisherman rows his boat through a *Gonyaulax* bloom, the sea lights up momentarily each time an oar blade is pulled through the water. If propulsion is by an outboard motor, the turn of the propeller sets up a helix of sparkles. Or even more spectacular, if you swim under-water through such a plankton bloom, as the water passes between your fingers, flashing speckles bounce off your face mask—psychedelics with-out drugs! In decades past, such luminescent displays were described as the "phosphorescence of the sea"—a spontaneous glow coming directly from disturbed seawater. We now know it is produced by living organ-isms, so it is described correctly as bioluminescence.

Gonyaulax

Even the collective light given off simultaneously by millions of bioluminescent alga is not bright. This being the case, it was reasoned that bioluminesce could not compete with sunlight and was therefore just not noticed during the day. The truth of the matter is that bioluminescent algae (like *Gonyaulax*) and animals (like fireflies) only produce light at night. This curiosity was discovered before the 1900s. A fruit jar was filled with seawater containing bioluminescent algae and shaken at various times during a 24-hour period: Light was emitted only at night. The same dedicated naturalist who made this observation carried his examination one step further. He took his jar into a closet and sat there in the dark for two straight days. At one-hour intervals he would shake the jug and record whether light was emitted. At the end of the second day a gnawing hunger drove him out of the closet and into the kitchen, but the experiment was at an end anyway because the algae were also starving in the absence of rejuvenating sunlight. Examining his records, the curious naturalist found that even in constant darkness the cells had only glowed at what would normally have been nighttime. He published his observation, but the idea that these microscopic algae possessed a clock never occurred to him; or if it did, he lacked the courage to make such a fanciful suggestion in writing.

But by 1957, when *Gonyaulax* had become readily available in pure culture, it was rather easy to carry out sophisticated studies of its ability to glow. Periodically air was bubbled through test-tube cultures of the alga bouncing the cells into each other and causing them to flash; any light given off was recorded automatically by a sensitive light meter. An obvious 24-hour nocturnal bioluminescent rhythm was found. Thus, each *Gonyaulax* cell, just like each *Euglena* and individual commuter diatom, possessed its own living clock.

## Life Without A Nucleus, Venus' Wine Glass

Here is another example from the world of single-celled plants. The very unusual Venus' Wine Glass anchors itself to the ocean bottom by a rootlike holdfast while the top of it gently waves in the massaging currents. During most of its life it is devoid of the upper umbrella-shaped canopy shown in the drawing; that structure forms only when the plant is about to reproduce. Amazingly, even though it may grow to five inches in length, it is a single cell! Because throughout most of its life it exists as just a slim filament sticking up from the ocean floor, few people ever notice it. But it is fairly common in warmer oceans; we collected our samples in the shallow waters off Mexico and Jamaica.[7]

Soon after collection, a laboratory that is lucky (mine) or skilled (all others) get Venus' Wine Glass into culture so there is always a ready

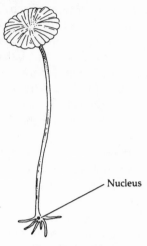

Nucleus

*Venus' Wine Glass*

supply available to use for experimentation. This plant does not glow in the dark or move around on its own accord, so the first rhythm looked for in it was one in photosynthesis. To do this, the photosynthetic rate was determined every few hours. This means that even though it might be the middle of the night and therefore dark, the darkness was interrupted by the same test-light intensity, and for the same short interval, as was used for daytime measurements. Thus, every determination was made under identical conditions. Nevertheless, the photosynthetic rate of the plant was considerably less at night than at day. Saying it another way, there would be no sense for the plant's photosynthetic machinery to gear up at night in nature because it can't function without light—so it does not.

Figure 5.1 shows a photosynthesis rhythm (as determined by the oxygen produced) during the first five days in day/night conditions. Though the results would be expected, returning to the figure you can see that after day five, the light was kept on all the time and the rhythm persisted. Thus, here is yet another organism that possesses its own clock.

As you know, the DNA-based genes of a cell nucleus rule most of the important events in each cell. In Venus' Wine Glass the nucleus resides in the holdfast part of the cell at the base. All a biologist has to do to remove it is cut off the holdfast. That drastic treatment sounds like an execution but it is not with this organism, the alga will continue to live for several months without its nucleus. As if that were not sufficiently startling, return to Figure 5.1 again and see that the nucleus of that plant

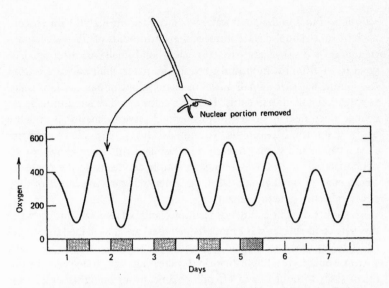

*Figure 5.1. The photosynthetic rhythm of Venus' Wine Glass, first in day/night conditions (days 1–5) and then with the light left on constantly. At the end of day 1 the nucleus was removed. The shaded areas on the abscissa represent 12-hour intervals of darkness.*

was removed at the beginning of the second day, yet the rhythm persisted even in constant conditions; it is known to endure for as long as 28 days! Thus this plant continues to tell time even when its nucleus— its master control center—is in the waste basket!

Before you catch your breath, there's more. In the next mutilation, an anucleate filament was cut into half-inch long segments. Each diminutive fragment continues to live for quite some time and the photosynthetic rhythms of each piece persist. The conclusion is inescapable: Each minute segment has at least one clock (and probably more) of its own! Venus' Wine Glass, in spite of being a single cell, contains many clocks, none of which require a master "Big Ben" in the nucleus to set the interval of their time signaling. A final show stopper is the fact that the photosynthesis rhythm is only one of many different kinds of rhythms in this plant.

## DAILY CHANGES IN SEX

I'll give one last example from the world of single-celled life—this time from a protozoan called *Paramecium multimicronucleatum*. Most readers will remember paramecium as a peanut-shaped, cilia-covered organism scooting around in the pond water you looked at through the microscope back in your school days. Paramecia have two kinds of nuclei, each distinguished by its size: large (macronucleus) and small (micronucleus). As you can tell from the mouthful-sized species name, *multimicronucleatum*, this animal has many small nuclei in its body. Sex in paramecia is called conjugation and goes like this: Two cells come together and mutually exchange micronuclei—which in this animal serve as sperm—through a bridge that forms between the two animals. This conjugal bliss completed, they separate and swim off, each now instilled with a new dollop of genetic material. Soon the exconjugates simply divide, as do the micronuclei, again and again giving rise to daughter cells carrying copies of their newly refreshed genetic makeup.

In studies designed to depose paramecian sex life, a curious aspect of this particular species was found: Its members are not divided into males and females; there are instead eight different sexes. While the politically correct would have a linguistic field day creating ways to describe the different sexes without fear of offending some, protozoologists simply distinguish between them by calling them mating type I, II, III, up to VIII. Additionally, the strain I am about to describe is both type III and IV and can change its mating type from one to the other without the aid of sex-change surgery and hormonal therapy used by some humans. In fact, sex reversal is originated via instructions emanating from a cell's clock. When maintained in a light-dark cycle of six hours of light alternating with 18 hours of darkness, this paramecium is mating type IV during the last few hours of light and at least through the first 12 hours of darkness. But then it converts to type III during the first hours of light.

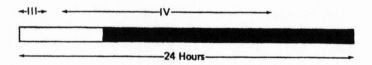

At other times of the day the animal is in sexual transition. As you must be getting used to by now, this sex-reversal rhythm, and like other rhythms will persist in constant conditions.

Here is a simple experiment that provides some very interesting, important, and surprising results. It begins with a large population of *Paramecium multimicronucleatum* that is kept in day/night conditions (as seen in the flask in the upper left-hand corner in Figure 5.2).

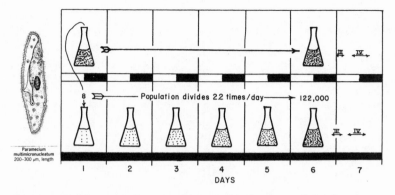

*Figure 5.2. The diagram outlines an experiment that demonstrated that the* Paramecium *clock kept on running even during cell division.*

Eight of these cells are removed and put into a small container of their own and kept in constant darkness (lower left-hand corner). Life is good in there because there is plenty of food in the culture medium, and they divide asexually several times a day; in fact in just six days these eight animalcules have given rise to over 120,000 new paramecia. Other than the eight original cells, the rest of their flask mates have never been exposed to a day/night alteration—instead, theirs has been an existence in eternal darkness. By day seven the food in their culture medium has been reduced greatly, and this paucity stimulates reproducing by sex (what a different turn-on from humans who often begin the rite with a pleasant meal in a nice restaurant): They begin their sex-change rhythm. Their rhythm is in almost the same phase as the rhythm of the large starting population from which the eight were culled. Now, to the fascinating conclusion: The clocks in those eight cells continued to run accurately, in the absence of any environmental time clues, while each *Paramecium,* and thus its clock, divided in two more than twice a day and still accurately retained its proper time. That is truly an amazing feat, being more impressive than a Timex watch "that can take a licking, and keep on ticking."

# RHYTHMS IN SHORE DWELLERS

## THE FIDDLER CRAB, A COMPLEX OF RHYTHMS

N ow let's leave the world of life in water droplets and turn to animals, all that can be seen with the naked eye. We started on the
seashore in the first chapter; now we return there again.

A rather common crab inhabiting the intertidal zone (that margin of
the shoreline that is washed by the tides, in most places twice per day) is
the fiddler crab (above). The males of the species have, on their attack
end, two claws: a major one that is as large as the body of the animal and
a tiny one. The large one is used for macho activities, while the puny one
is used like chopsticks for eating. Many years ago, a sun-crazed naturalist

lyrically identified the big claw as a violin, thus the name fiddler crabs (except on Bora Bora in French Polynesia where they are called violin crabs). Considering the actual morphology of the big claw, nutcracker crab might have been a more appropriate appellation, but I suppose that even with crabs, mien is in the eye of the beholder.

Speaking from an evolutionary point of view, fiddler crabs are quite successful: They are found on shorelines all around the world; the ones I study are near the Marine Biological Lab on Cape Cod, Massachusetts. Most of them have bodies not larger than about an inch across, and they come in a variety of colors ranging from black to olive-drab to brilliant reds, yellows, and blues, depending on the species.

A typical day in the life of a fiddler crab consists of sitting out each high tide underwater in a self-dug burrow and then emerging out onto the exposed mud or sand flats to scurry around (they run sideways) during each low tide. Because they can live both in air and underwater, they are described as being semiterrestrial. The above-ground routine followed by the males has been described as the 4 F's: feeding, fossicking, fighting, and mating. Life is relatively harder for the males than the females since they have only one claw to use in feeding. The major claw is so long, and the arm it is mounted on so short, that food held in the claw's tip cannot be brought to his mouth. A male also requires more food than a female because in essence he is eating for two: himself and his enormous pincer. Additionally, mealtimes are often interrupted to carry out ritual struggles with other males; these lead to nothing since pinching does little damage to a combatant dressed in a panoply of skeleton plates. (But they try anyway: When you put a rambunctious lot of them together in a bucket they enjoy group-grope wrestling and all become clamped together in one united tangled mass, joined by their giant claws.) The fiddle claw is used also to thump on the ground and to wave suggestively to passing females (there is no spontaneous carnality in this species, females must be enticed into bachelor burrows). The latter dalliance sometimes ends in a successful pick-up that is followed by a two-day underground love-in. Roughly 14 days later the female emerges from her digs and ambles down to the water's edge. Under her abdomen, attached to rows of swimming appendages, are her offspring, developed as far as tiny larvae but each still enclosed in a membrane. It is nighttime, and the moon is usually at its new or full phases when she backs into the water and begins to splash her abdomen vigorously. This action causes the membranes covering the larvae to rupture and in the next ten minutes she becomes the proud parent of up to 90,000 young . . . after which, figuring that her procreative efforts have already been more than magnanimous, she indifferently abandons her offspring to fend for themselves.[8]

During high tides the crabs sit deep in their burrows. Some crabs even plug their burrow entrances with a glob of mud. When the tide ebbs and exposes the sea bottom, the burrows remain flooded for a considerable period of time, the length depending on the porosity of the soil and the underground drainage rate. But the crabs emerge soon after, and sometimes even slightly before, the sea floor is exposed to air. How do the still inundated crabs know when to emerge? A lab experiment reveals the means.

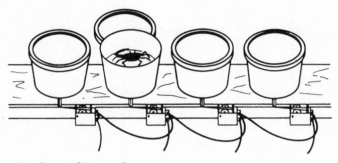

*Figure 6.1. Carrousel Actographs*

For a human willing to tough out an occasional puncture wound inflicted by a male giant pincer, it is easy to quickly collect a goodly number of fiddler crabs for study in the laboratory. There they are put into individual carrousel actographs (that is scientific bombast for what are really used margarine tubs) (Figure 6.1), which are attached to a machine that measures automatically whether they are traipsing in circles or standing motionless in their tubs. The actographs are then placed in an incubator from which all the obvious time and tide cues have been purged. So housed, when the tide recedes in the crabs' former kingdom on the shore, and the free crabs emerge and begin their 4 F-ing, the animals in the incubator run in circles; and when the tide re-floods their old home area they mysteriously (for caged wild animals) stand motionless—their version in the laboratory of being home in the confines of their flooded burrows. The pattern will persist for days to weeks and is called a persistent tidal activity rhythm (shown in Figure 6.2). Note that the twice-daily activity peaks come somewhat later each day, just as the tides do in nature. Clearly, in the laboratory fiddlers' built-in clocks take over the control of their temporal lives. The display portrayed Figure 6.2 is a very fine one: Not all crab rhythms are this clear and precise. When we stumble onto a virtuoso crab like this one, the

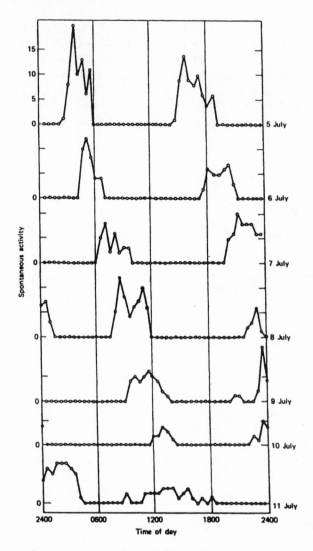

*Figure 6.2. The activity pattern of a single fiddler crab in constant conditions in the laboratory.*

crew of students in my lab feels the performer deserves special recognition. Thus, keeping with today's craze for hyperbole, such as calling championship football contests the superbowl, and cheap airline tickets supersavers, my students/comedians call him supercrab (or as a result of the English influence on our public "telly" stations: fab crab, the nom de hot-shot I favor). This crab's rhythm persisted for days in the atidal laboratory, but then we intentionally halted this wonderful cadenza to release him (still rhythmic) back into his home territory—hopefully to spread his fine tide-pool talent genes widely. These releases are done without complaint from my eternally-hungry students; they do not fancy fiddler crab *du jour*—but lobsters are a different story! When finished with an experimental subject, Charles Darwin used to say, "It's still good for the pot."

Just under their shells, crabs have specialized dermal cells called chromatophores (color-bearing cells), that are shaped somewhat like a seagull-dropping splatted on your car's windshield. The chromatophores contain dark pigment granules that are either concentrated in a cell's center or distributed evenly throughout the cell's radiating arms. When the pigment in all an animal's chromatophores is dispersed in the cytoplasm, the overall appearance of the animal becomes a Darth Vader black; when the pigment becomes concentrated into a tiny dot in cells' centers, the animal's suit of armor blanches. In their natural setting on the shoreline, when the crabs emerge from their burrows during daytime low tides their shells are black, but when they venture forth at night, their shells have lightened. As with the activity rhythm described in the previous paragraph, when this color-change rhythm is studied in constant conditions in the laboratory it too persists. The difference between the two rhythms is that the successive peaks

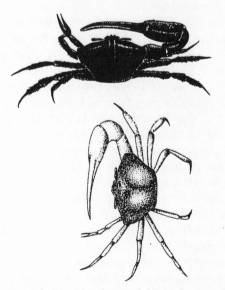

*Figure 6.3. Daytime (above) and nighttime (below) coloration of fiddler crabs.*

of the activity rhythm form at near 12.4-hour intervals (it is a tidal rhythm), while color change is a daily rhythm in which peaks are separated by roughly 24 hours. Thus, in analogy, this crab's living clock is like a human surf-fishermen's watch that tells both the time of day and state of the tide.

But a component of the color-change rhythm also cycles with a low-amplitude tidal frequency. And this rhythm has been found to be expressed even in parts of crabs isolated from the rest of the animal. The story goes like this. A fiddler crab has eight legs that it uses for walking (the remaining two terminate in pincers);they pretty much stick out in all directions from its body. Thus, if a predator, say a gull, grabs a crab, it usually gets it by one of its handy handles, a leg, which is torn off. While the gull snacks on the sacrificed hors-d'oeuvre appendage, the now seven-legger crab galumphs off to safety in its underground burrow.

So much for background. Crabs are brought into the lab, and a pair of forceps are used to fake a gull's bite: A hard pinch on a leg and off it comes. The loose leg is then put into constant conditions and examined at régular intervals for color changes. The rhythm persists as long as the leg remains alive!

## GREEN CRABS

The green crabs lives side by side with fiddler crabs on America's northeast coast. (The green crab is sometimes red, demonstrating one of the many weaknesses of using common names for animals.) It also displays a daily rhythm in color change just like the fiddler crab. Another similarity is that it has a clock that produces a tidal activity rhythm, but one that differs from the fiddler crab's in that the animal is active during high tides, *meaning that it runs around under water rather than when the tide is out.* As a result of this difference in phase, the two crabs can live contiguously and do not have to openly compete for food at the same time.

Green crabs also live on the shoreline of the Mediterranean where the tidal exchange is very small. Laboratory studies were carried out at the Stazione Zoologica Napoli on the activity rhythms of local green crabs. No sign of a persistent tidal rhythm was found in the Neapolitan crabs, instead their activity rhythm was of the one-peak-per-day variety. This suggests great versatility in these crabs. It would be informative to transport the crabs to the English shoreline, expose them to the two-tides-per-day regimen for several days, and then test to see if they would express this rhythm in the laboratory.

Considering that green crab activity is daily in the Mediterranean and tidal elsewhere, it would be only normal if readers assumed that plants and animals somehow *learned* to be rhythmic by exposure to environmental cycles like day alternating with night and tides flip-flopping between high and low. As intuitive as this conclusion may seem, it is not correct. We are all born with our clocks, but, depending on the species, some need to be started up, while others hit the ground running. The green crab's clock provides one of the many proofs of this, as I will now describe.

A fine English marine biologist Barbara Williams, who intrepidly tackles Sisyphean tasks, chose the exceedingly difficult chore of becoming a surrogate mother to a batch of green crab eggs. Having a zoologist's green thumb she nurtured the crabs to near adulthood in her New Zealand laboratory. When they finally reached a size suitable for a few days' stay in an actograph, in they went. Yet, over the next few days, they showed not the slightest hint of tide-mimicking activity.

Next Dr. Williams did a curious thing: She submerged the animals in near freezing water for several hours, warmed them up again, and put them back into the actographs. And while the crabs had never even seen the ocean and could have not learned anything about the tidal pattern from a single dunking in ice water, low and behold, their activity pattern thereafter assumed a tidal rhythm! The interpretation is that their body clocks were there all the time, but just had not begun working until subjected to a cold shock. It was just the same as a spring-loaded alarm clock that you must wind up before it starts working. Or, if you prefer a living example: like slapping a baby's bottom at birth to start its rhythmic breathing movements. The carry-home message here is that the living clock in innate—it is part of our genetic makeup; rhythms are not learned from the environment after we emerge into it. Setting the phase of tidal rhythms as you know from the discussion of jet lag in Chapter 4, environmental cycles do play an important role in setting the hands of the living clock.

Those of you who have had the pleasure of traveling from cove to cove along a few miles of shoreline may have noticed that the tide can be in quite different stages of flood or ebb. This disharmony is due to the different contours of the shoreline and shape of the local ocean bottom. Should one examine the setting of crabs' rhythms in each cove, it would be found that they are in tune with the tides of their own cove. Experiments have been designed around this phase-setting ability of the tides; here is a fine example of one. There is only one tidal exchange each day on the Caribbean shoreline of Costa Rica, while on the Pacific side the tides occur twice daily. Fiddler crabs collected from the former display only one activity peak per day in the laboratory, while those from the west coast show two. When crabs were transported overland from the east coast and exposed in cages to evocative Pacific tides for a few days, their rhythms showed that they had adopted the twice-per-day tidal schedule.

## The Clockle's Clock

Living buried just under surface sands in a beautiful bay on the South Island of New Zealand is a cockle, which we affectionately renamed, metonymically, the clockle. During low tides the sand overhead (I guess immediately above would be better since clams do not have heads) is exposed to air and the clockle closes its shells tightly—it clams up. During high tides it opens its shells just wide enough so that water can be pumped into its mantle cavity where microorganisms are filtered out as a source of food. The opening-and-closing ritual is called a shell-gaping rhythm and will persist when the animal is studied in the lab away from the tides.

## Clams As Calendars and Historians

Each time a clockle opens its shell, a bit of new calcium is deposited on the shell margins; in this way all clams grow in size. When the shells are closed together, a dark material is laid down. Thus the shell is visibly layered. When a shell is cut crosswise and the cut edge polished smooth, using magnification one can see the individual layers that have been deposited during low-tide intervals. Thus, just by looking at the shell in this manner, one can know exactly what the monthly tidal schedule had been until the clockle was placed in the collecting bucket. This simple way of studying history has some far-reaching consequences.

As the Earth ages, its rotation rate slows down. One of several ways of establishing this as a fact is to use fossil clam shells, or even better, fossilized corals; both serve well as silent witnesses to history. Here I'll focus on corals as they are easier to interpret. As they grow taller they lay down one layer of calcium each day, and the thickness of each layer is a function of the season of the year. Thus, looking at a section of a modern coral you see thick layers becoming gradually thinner and then enlarging back to the original thickness. And when you count the number of layers within this area, you find there are 365, one for each day of the year. The pattern makes coral stems useful as simple, rather unambiguous, natural calendars, and they are thus described as being geochronometers. With this fact established, one is in a position to examine the passage of time during the past lives of now-fossil corals, and we find that they can serve as paleontological clocks. Suppose we looked at one that lived during the Devonian Era that began 405 million years ago. Counting layers laid down during this era you learn that there were 400 days in a Devonian year, meaning that the earth was spinning faster then—days were 22 hours in length!

Here is another way of studying history. On some shorelines, where conditions are just right, each flood tide deposits a layer of mud and sand on the shoreline. With time these layers become compressed and eventually are transformed into clearly layered sedimentary rock. These rocks are called tidalites, and ancient ones, ranging in age from 305 to 900 million years old, have been collected from old, uplifted sea bottoms now called Indiana, Alabama, Utah, and Australia. The deposition layers within them vary rhythmically from thick to thin, depending on whether they were laid down by the spring or neap tides. The alternating lamina of tidalites formed 900 million years ago indicate that the Earth then revolved on its axis 30 percent faster than today, meaning that the year had 481 days, each 18.2 hours long!

At present the Earth is losing about 2 seconds every 100,000 years, meaning that finally we have the answer to, "where does the time go?" An especially valuable aspect of these approaches to history is that revisionist historians cannot distort these data—they are set in stone.

Here is a real show stopper: The actions of humans are actually reducing the rate at which the earth is slowing down. This fact has been documented by Dr. Benjamin Fong Chao, a geophysicist at the Goddard Space Center. Most of the water-storage reservoirs we humans have built are in the Northern Hemisphere, meaning that we have concentrated the water's weight closer to the earth's axis than if the reservoirs were located on the equator. The impounded water's combined weight is close to a trillion tons. All this weight moved closer to the earth's axis causes our planet to rotate faster (just as figure skaters can make themselves spin faster by drawing their arms into their bodies). As one would expect, the reduction is minuscule, but can be measured by Dr. Chao's instruments.

## EARTH TIDES

Here are a few words on how tides are created. As the sun, and especially the moon, pass overhead (more accurately said—thanks to Copernicus and Galileo—we know it is the Earth rotating under these heavenly bodies) their gravitational forces pull the Earth's oceans below them upward, creating a high tide (for this discussion I need not go into the generation of the tide that forms simultaneously on the opposite side of the earth). What most people are unaware of is that these forces also raise the solid ground under our feet—creating a dirt and rock earth tide! For instance, as the moon passed over what was the World Trade Center in New York its buildings rose approximately fourteen inches, and when the moon first rose in the east the buildings tipped two inches toward it, and when it set they leaned two inches toward the west. Those of us standing in buildings slowly go up and down in synchrony with the Earth but we are oblivious to the fact the we too (like the Chapter 1 organisms) undergo a vertical-migration rhythm! A more dramatic example is the Earth tide on Io, one of the four Galilean moons of Jupiter. It is less than a third of the size of Earth but its solid surface rises and falls in 300-foot tides. (Isn't it amazing that we can know something like that is going on 392 million miles from Earth!)

With that as background, here is an interesting story involving the Earth tide that raises some big questions—but does not answer them—about the reality of the living clockworks. There is an insect erroneously called the cave cricket which is really a grasshopper and doesn't always live in caves. Taxonomists can be ruthless with each other when fighting over the correct name to assign to a species; and when I first worked with this insect it was called, somewhat redundantly, but romantically, *Hadenoecus subterraneus*—the insect from the dark bowels of Hell. Its new modern name is more pedestrian and less intriguing. Anyway, a batch of cave crickets were confined in actographs and placed in natural constant conditions: They were sealed off deep in an old mine shaft shuttered with heavy steel doors. In there the air temperature held constant at 40°F, the humidity at 93 percent, and the darkness was of the kind only understood by spelunkers when their flashlight batteries die. In the cave with them were sensitive strainmeters designed to measure Earth-tide alternations. The steel doors were opened only every one to three months for a brief inspection of the equipment. When Ruth Simon, the investigator on this project, analyzed the activity rhythms of the cave crickets, she found a strong positive correlation with the Earth-tide rhythm! Could this have been a cause-and-effect relationship? Should we now consider standard constant conditions used in the laboratory not to be interpreted as constant by our test organisms? Are they responding to the Earth tides? To the best of my knowledge, this work has never been confirmed by other workers, so it is too soon to even ask these questions.

Here is another curious example of a correlation between biological activity and Earth tides that comes from the pen of Dr. Ernst Zürcher in Florence, Italy. He and his colleagues placed two, five-foot-tall spruce trees, growing in separate pots, in a dark room in which the temperature was held quite constant. Every few hours they measured the diameters of the stems and found that changes were expressed as a tidal rhythm! The rhythms of both plants were almost identical, other than a difference in the amplitude of the curves. The investigators compared these rhythms to the form of the Earth-tide rhythm on each day and found a near perfect inverse relationship: When the stems swelled maximally, the Earth-tide rhythm troughed. This relationship held even in sections of stem isolated from the plants as long as they remained alive. Just how this apparent relationship is brought about is unknown; it seems highly unlikely that gravitational forces on the spruce could physically produce the changes.

One must be circumspect in interpreting the findings reported in these last two studies. What was discovered—stated in statistical parlance—is a correlation, a condition where two processes vary in a way that suggests they are truly associated. There was a positive correlation between cricket activity and earth tides and a negative correlation in the stem-diameter study. The most optimistic conclusion that can be made in either case is that the correlations described *may* be cause-and-effect relationships. One is always tempted to conclude that a relationship is causal, but that is very often not the case. For instance suppose someone noticed that all the people who work eight-hour shifts die. While the correlation is strongly positive, the correct conclusion is obviously not that such a workday causes death. We all die, and most all of us employed have a normal work load; the relationship is just happenstance. On the other hand, premature death does correlate meaningfully with cigarette smoking.

## What Samoans do at Thanksgiving

Here is an example of a daily, a monthly, and an annual rhythm, all wrapped together and displayed in a single species, the Samoan Palolo worm. This aquatic relative of the earthworm has a mating ritual more steamy than those in a soft-porn paperback. The worm lives in tunnels in the near-shore reefs off the Samoan and other islands of the South Pacific. At the end of summer, an extension grows out of each Palolo's tail end,

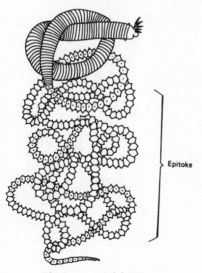

Epitoke

*The Samoan Palolo Worm*

called an epitoke, that consists of lots of individual segments all filled with either green eggs or blue sperm, depending on the sex of the *derrière* that spawned them. When these new aft ends reach maturity the worms engage in an annual orgy. At dawn, mainly on the day when the moon reaches its last-quarter phase, in November, the sexual appetency of the foot-long, tail-shaped buttocks has matured, and they break free from their creators and swim to the ocean surface. There they begin a decerebrate, hootchy-kootchy tutti that churns the top waters into something akin to a frothy brothel. Native Samoans stand ready in chest-deep water and in canoes until a lookout sees the rise and gives the clarion call, "*Ua sau le Palolo,*" which translates roughly to "scoop up the wrigglers before they explode." Buckets full of the worms are caught for a subsequent feast, but some vermes-loving fanatics cannot wait and gobble them down on the spot. Disciplined gourmands stifle their desire long enough to get the worms home so as to dine on delicious roasted Palolos that turn emerald

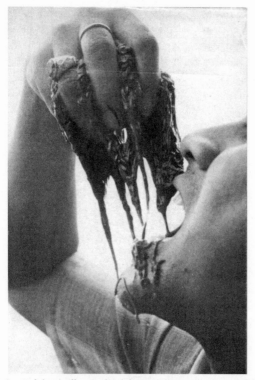

*A handful of living Palolos (still wriggling) being consumed* au naturel *by a* vermes aficionado.

green over the fire. This takes place at roughly the same time as we in the United States turn Thanksgiving turkeys into carcasses. To us, worm-shaped animals and our untested impressions of what it must be like to eat them are turnoffs, but in truth Palolos are low in fat and higher in vitamin A and carotene than chicken eggs. In other words, they may be better for one than the junk food many of us cherish. Incidentally, if you wish to order Palolo when dining in Fiji, ask for *mbaloba*.

But I digressed (it was just before lunch when I wrote the above) . . . we must return to those lucky worm backsides that manage to escape selfish human spoil-sports who put their own gastronomic delights before a Palolo's sexual gratification. We left the worms writhing on the ocean surface. Almost in synchrony the wrigglers explode, die, and sink to the bottom, filling the sea with technicolor eggs and sperm during their big bang. The opposite gametes fuse in enormous numbers and the species has been perpetuated by adults substituting death by explosion for orgiastic fruition.

## THE MOON AND HUMAN SEX LIFE

There are many examples of organisms synchronizing their sexual dalliances with a particular phase of the moon. Could humans be among their ranks? Certainly some people adhere to this notion, and we read and hear in popular literature, poetry (moon, spoon, June), and song (the haunting: "When the moon hits your eye, like a big pizza pie, that's amore") that the moon must directly influence some aspects of our lives, especially sex (and sex crimes if you ask certain criminologists). To try to shed some light on this question I teamed up with Richard Udry and Naomi Morris at the Carolina Population Center to produce the following study. Seventy-eight couples (all married I might add—this study was done quite a while ago) were paid a dollar a day (change that to lots of years ago) to report, via post cards, whether they had sex in the previous 24 hours, and if so, at what time. The moment of orgasm is not a good time to make a decision (proof of that comes from people who practice *coitus interruptus* as a mean of birth control: many are now called parents), so the husband and wife were asked not to consult with each other while making out their postcards. But the agreement in the times they reported was so remarkably similar it made one wonder if the copulants were truly concentrating on performance, or just clock-watching. However, their answers made it quite clear that they, unlike President Clinton, knew the exact definition of the word sex: both partners had it. At the end of the study we had gathered 5,584 of what we called copulating-couple days of data. The first thing we searched for in our new databank was the most popular time of day for having intercourse. Before I give you the answer, why not take a guess?

The result of our study are seen in Figure 6.4. Examine the curve built of open triangles. Fifty-six percent of daily copulations took place between 11 P.M. and 1 A.M. There is also a secondary hump (if you will) in the data at 7 A.M.; a time of day when a man's blood testosterone (a libido-building hormone) level is highest for the day.

But the point here was to determine what role, if any, either directly or indirectly, the moon played in this timing. First, some background. The interval between successive moon rises is not 24 hours, because, unlike the sun, which is a relatively stationary heavenly object, the moon is not. It orbits around the Earth, while the Earth simultaneously rotates under the moon. To catch up with the moon's orbiting, the Earth must turn for an additional 50 minutes, thus the intervening period between successive moon rises is 24 hours and 50 minutes, the lunar day. To construct a curve like I show for the daily rhythm, I used hourly averages of the 24-hour day. To construct a curve for the lunar day I simply used lunar hours of the 24-hour 50-minute lunar-day. If done with the proper number of

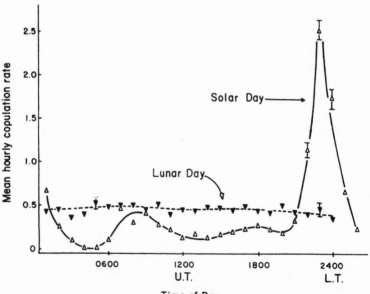

Figure 6.4. The curve marked by open triangles represents the average times at which 78 couples copulated during the day over an interval of several months. The favored time is clearly between 10 P.M. and midnight. The same data were when replotted according to the hours of the lunar day. The curve (solid points) produced is a straight line indicating no correlation with the moon.

lunar days, the solar day influence is mathematically randomized and disappears from the data set. If the moon had any influence on couples' sex lives it would appear as a peak, or peaks, on the curve, otherwise an essentially straight line would be the result. It was the latter that we found (the dashed line based on the inverted solid triangles).

We also looked at the data in several other ways. Because our subjects were working couples, we examined the Sunday copulation pattern when sexual urges would be less likely to be repressed for out-of-the-house chores and obligations. Again, we found no association with the moon, but I can relate with considerable authority that few of the subjects went to church or out on the golf links in the morning, and that an alternate way had been found to fill the otherwise wasteland of a Sunday afternoon . . . but still the major copulatory peak occurred around midnight. However, the study did not confirm the American writer Susan Ertz's concern that "Millions long for immortality who do not know what to do with themselves on a rainy Sunday afternoon."

We then compared the number of copulations when the moon was above the Earth's horizon with the number when it was below (when "moonbeams" would be blocked by the bulk of the earth)—and also the number that occurred at the major phases of the moon over a month—but found no significant correlations. So, I must leave the subject with the conclusion that there is no man in the moon (or man and women for those who see in that construction social benefits sufficiently great to outweigh its awkwardness) overseeing and directing human sex life.

Now, for the sake of landlubbers, we will leave the sea and move inland.

# SOME ANIMAL RHYTHMS

## A USEFUL PEST, THE FRUIT FLY

Before science corrected a seventeenth-century interpretation of an everyday observation, people, using their eyes and noses, adopted one of the more believable hypotheses of the day: Life arose spontaneously from decaying meat. This seemed obvious: Meat left on the kitchen counter, or even more conspicuously if left outside in the sun, was soon teeming with disgusting white crawling worms (actually fly larvae). The conventional wisdom of seeing is believing (a sententious dictum that usually makes good sense) produced the spontaneous generation hypothesis. i.e. life arising from nothingness However, Francesco Redi destroyed this verisimilitude by performing one of the most inexpensive experiments ever done in the history of science: All it required was a piece of almost spoiled meat placed in a mayonnaise jar whose opening was covered with a cheesecloth rag. He left this scientific apparatus sitting in his lab where the meat soon stunk to the high heavens, but no white worms appeared on it. Instead, the worms arose spontaneously on the diaphanous cover. Wisely, Redi did not conclude that all life arose from cheesecloth! What had happened was that flies could not reach the malodorous meat in the jar so in desperation they just laid their eggs on the protective netting cap.

Nowadays, with refrigeration, and meat being the most expensive item in our food budgets, we seldom let it go bad; but we still often have the opportunity to see a modern-day version of food-into-flies creation. After a bowl of fruit has sat for a few days on the counter, a nimbus of fly-speck-sized flies too often mysteriously begin circling it. While cursing, you quickly seek out the one spoiled pear, apple, or whatever; and if you look closely will see the tiny eggs on the fruit—progenitors of the soon-to-be winged version of the fruit fly.[9]

Emotional evaluation of the existence of this pest range from the insecticidal cry to call out the SWAT team to recognizing it as one of the most important animals in the genetics lab. The reason for the positive end of this range of assessments has to do with the ease of culturing the fly in the lab and the fact that it can produce a new generation in as short a time as just two weeks. Additionally, they are prodigious breeding machines: A single female lays hundreds of eggs during her lifetime. And, because of their tiny size (only three millimeters long), large numbers can live their entire lives in a half-pint milk bottle—their traditional laboratory home.

The fly's life cycle goes like this. Eggs quickly hatch out worm-like larvae, and after these feed and grow for a few days they secrete a sarcophagus-like case around themselves and enter what is called the pupal stage. Within the hard pupal case a miraculous transformation—a small-scale version of the ugly duckling story—takes place: The lowly larva transmogrifies itself into a six-legged flying machine. When the transformation is complete the pupal shell splits down the back and the adult fly, wings still folded around its body, wriggles out. The process of shedding the old covering is called ecdysis, a word H. L. Mencken used to construct "ecdysiast," his hyporcorism for a stripteaser. (He also coined "bibliobibuli" for people who read too much, but don't let this mention stop you now). So that there be no confusion between fly and dancer here I will use another term, "eclosion," which definitely refers only to insects. To carry on, once the adult has emerged completely, it pumps blood into its folded wings and inflates them to full flapability. The wings must then dry out and harden before the animal can fly.

In nature the adults eclose, i.e., emerge from their pupal cases, at dawn. This scheduling is very important because at the first light of day the atmosphere is most humid. Thus, in the damp air, while the fly is struggling to expand its wings, they will not dry out and harden unexpanded and nonfunctional. Additionally, at this moment the fly's cuticular body covering is still quite permeable, therefore body water can be lost twice as fast through evaporation as it can a few hours later when the covering has hardened. The insect's scientific name is *Drosophila*, which translates to lover of dew.

Thus, in nature a fruit fly population has a prominent eclosion rhythm—with emergence peaks coming each morning—that is easy to study in the laboratory. When a quantitative examination of this rhythm was first attempted, undergraduate students served as data collectors by manning bang bottles. These old milk containers had funnels for caps (worn like the Tin Man's chapeau in the Land of Oz), and at one-hour intervals, the students would tip a bottle of flies over and bang on its bottom. Any flies that had emerged from their pupal cases in the previous 60 minutes were knocked out into a dish of detergent where they drowned, and their numbers were determined. (Later, human labor was replaced by an automatic shaker and fly counter.) Needless to say, the students found that if the flies were kept in normal day/night conditions maximum emergence occurred just after dawn. Next, the pupae and students were moved to a closet so counts could be made in constant darkness (the very first time this rhythm was examined, it was done in an outhouse [a one-holer] in Colorado). The rhythm was found to persist, meaning eclosion was clock controlled. Because the larvae and pupae developed at different rates, they did not all eclose on the same morning. Slower developers emerged on subsequent days, but when they did arise as adults it was always within a six-hour interval close to what would be the time of dawn. This means that flies developmentally ready to eclose in the afternoon or at night were forbidden to do so by a very restrictive clock acting as a gate; the flies' emergence had to wait until the next morning.

The flies used in the experiment described above were allowed to develop from egg, through larva, and pupal stages while maintained in day/night schedule before being studied in constant conditions. But if those developmental stages were forced to come about in constant darkness, the adults eclosed at random, showing no sign of a rhythm. Just to see what would happen, 15 consecutive generations of flies were raised in constant darkness, during which they remained arrhythmic. Then a very simple thing was done, the light in their culture room was turned on and left on. From that instant onwards, the population was rhythmic! In the next, slightly modified experiment, the light remained on for only 1/1000th of a second. Produced by a strobe light, this microswitch burst of light was sufficient to initiate the eclosion rhythm that then persisted in the ensuing eternal darkness. Clearly no fly could learn anything about the 24-hour interval of a day by either of these treatments; the conclusion must be—as it was for the green crab shocked by a cold pulse (Chapter 6)—that the clock was present all the time, and these simple treatments just ignited it. This was another proof that rhythms are innate, not learned; the clock is built right into the genetic material of the fruit fly—and the crab, and us, and all other living things.

If the clock has a genetic foundation we should be able to manipulate it. The type of manipulation I will describe here is done all the time in agriculture. For example, if a dairy farmer finds that one of his cows gives significantly more milk than the others, he will breed her in hopes that the trait will be passed on to her offspring. When the trait does pass, he continues the mating ritual and in the end produces a herd of superb milk producers. The same trick was tried with fruit flies. Most of the flies in a culture eclose at first dawn, but a few, a tiny slugabed group, emerge several hours later. The early flies were mated only to each other; as were the late emergers, and their offspring were treated similarly. This continuous ritual of mating offspring-earlies to other earlies, and lates to other lates was repeated 16 times. Had this been an experiment using cows it would have taken years, but because of the short generation time of the flies it took only weeks. The result was the production of two inbred populations, one that emerged at the normal time (dawn) and another entire population that eclosed later in the day. Clearly, the time of peak emergence of the eclosion rhythm is under genetic control. The mysterious clock we are looking for, thanks to its genetic basis, can be manipulated at the will of an investigator. As will be described in the last chapter, the period (the time interval between peaks) of the rhythm is also under genetic control.

*Drosophila* also display clock-controlled activity rhythms. Like most mammals and birds they are active during daylight and inactive at night. Mammals and birds sleep at night; do fruit flies? Here is some evidence. Neurophysiologists just reported that if flies who have just recently become inactive at night are subjected to very slight vibrations, they are quickly aroused to activity. However, if they have been inactive for a while, vibrations up to 120 times stronger are required to rally them. This is a slumber pattern strikingly similar to that of humans and mammals. In humans, caffeine increases alertness on waking, while many antihistamines cause us to become sleepy. *Drosophila* responds the same way to these substances. Finally, there are certain genes in mammals that turn off during sleep and on when awake; the activity patterns of the same genes in the fruit fly are equivalent. It appears that the little buggers actually do sleep at night.

We'll leave the fruit fly here with one last interesting observation: The 24-hour day length is very important to this insect. Just to see what would happen, the fly was kept in two different, artificial day lengths: one whose light and dark intervals totalled 27 hours, and the other that totalled 21 hours. The outcome was a significantly shortened life span compared to the control animals that were maintained in a 24 hour day. This reduction in longevity is a common finding in other animals also when they are exposed to non–24-hour light/dark cycles. Numerologists take note.

## THE EARLY BIRD VS THE LATE WORM

I will begin a discussion of bird clocks with a story about worms. As a kid I helped support my family by fishing a trotline in the Mississippi River and selling my catch at riverside to passing motorists. Being ten years old I always sold out before the adult fish mongers who were not at all pleased with my cute presence and competition—in spite of the fact that I fixed my prices to their fixed prices. During the rare times any of them deigned to speak to me, the subject matter was always twofold: What did I use for bait, and didn't I realize that they had to support families on their earnings, while I was only playing. To the latter I rebutted simply that they were better dressed than I (back then I was shabby respectable), and to the former I revealed that I used worms. (a useful tip; all I learned from them was four-letter words). "Ha, the kid uses worms!" They each had secret formulas for their baits. They boasted that that's why they caught more fish than I (which was true; and now that I own probably a thousand dollars worth of fishing equipment I catch even fewer fish), plus they didn't have to dig worms. I didn't tell them, but I didn't dig either: I learned that I could go out after dark and pick worms up off the ground.

Nightcrawlers, the mega-sized earthworms I gathered, have clocks and undergo a kind of vertical-migration rhythm: During the daylight hours they stay underground, but at night when they are not feeding they rise and extend their anterior ends out of their burrows. They keep their posterior ends firmly anchored to the walls of their burrow, so that should the vulnerable exposed front end be tormented, powerful longitudinal muscles running the entire length of their bodies all would contract simultaneously, yanking the front end back into the safety of the burrow. By timing their excursions in to the night they remained relatively unbothered while they gulped down dead leaves and grass. I took advantage of their nighttime emergence: Using a red light, I crawled around the yard popping them into a tin can before they knew what manhandled them out of house and hole. Occasionally I encountered two worms side by side, and when I snatched both up with a single grab, I would call out enthusiastically to the empty darkness, "double," not realizing I had just destroyed the sadomasochistic (they drive sharp spikes into each other's bodies to hold themselves in the proper mating alignment) conjugal bliss of those domestic partners (the books for ten-year-olds in those days did not mention sex; and I had not yet read Charles Darwin's 1891 treatise on earthworms, or *Lady Chatterley's Lover*). Lest it be lost in the nuttiness of this digression, the point here is that the worm clock is very adaptive in that it adjusts the feeding and reproductive efforts of the vertical-migration rhythm to the nighttime when few predators are about, and the cool moist air protects the worms from excessive water loss through their thin skins (and thus death). Having a clock is highly adaptive because it saves nightcrawler lives and enhances the perpetuation of the species. But what if the clock turns inaccurate, or a worm overrides the dictates of its timepiece and stays on the surface too long? Simple: The early bird gets the worm (this aphoristic wisdom does not apply to all species, for example, it is the second mouse that gets the cheese). Here then, is a war between biological clocks. Bad worm clocks lose out to pushy, punctual robin chronometers.

Finally, on to the study of bird activity in the laboratory. Birds are much better subjects for study than worms because worms are so sluggish that one has to wait a long time for them to do something. And most people are more enamored with birds than bird food. As a result, the literature is rife with bird studies. All kinds of species have been used: game birds, barnyard birds, songbirds, seabirds, et cetera. Even I have dabbled with our feathered-friends' rhythms; Figure 7.1 shows the result of one such study.

When I was a graduate student I happened to trap some birds in the spring that were migrating north to breed. These were perching birds that migrate at night and feed during the daytime. I put them into a recording apparatus and discovered that while they had nice daily

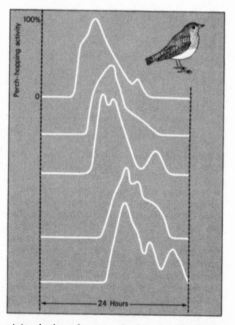

Figure 7.1. The activity rhythm of one caged robin, over five days, in constant conditions. The rhythm persisted nicely, but the period—the interval between successive major peaks—lengthened slightly (to 25.3 hours) from the 24-hour period displayed by the bird in its natural habitat.

rhythms as expected, but these also displayed a second activity bout at night. They did this night after night even though they were in the lab and thus didn't know it was dark outside. Clearly this nocturnal activity was clock generated, and I was quite excited about making this discovery. I rushed into my mentor's office to tell him of my great find. He tactfully broke it to me that German biologists had discovered migratory nighttime activity some 100 years ago and named it *Zugunruhe*... I had begun my scientific career by being scooped a century ago.

Birds' clocks do much more than just govern daily activity patterns, and some of the behavior they are involved in is surprising. For example, avian chronometers are involved in homing pigeons' ability to find their way home. A standard experimental procedure is to load birds into a car, drive them miles from their loft and release them. Then, you turn your car around and race back to the loft to see who gets there first (a few birds never come home, but good riddance, they would just sully future experiments anyway). On release, unless you try to follow pigeons in a plane, all one can do is jot down the compass direction they were heading when they vanished from sight. But if you measure the time it takes for them to reach the home loft, you find they fly in a pretty straight line. How do they do it?

The problem with this kind of field study is that there is no way to control it, and you can only watch what the birds do for a couple of minutes until they disappear from view in the distance. Thus the study was brought indoors. A circular cage/feeding tray was built with eight or more identical feeding stations evenly placed around the circumference, and a stationary, artificial sun placed off to one side. At 9 A.M. each day, the south feeding station was provided with bird seed (in a way such that a bird, at its release point in the apparatus's center, could not see the food), and a bird was released. During the first trials the bird had to use a random search to find the station containing food, but it rather quickly learned to go directly to the south food station. In doing this, how did it find the food, since all the stations were overtly identical in appearance? The only landmark in the whole room was the artificial sun, so it was suspect. If a bird released at 6 A.M. simply kept the sun to its left side—formed a 90° angle with the sun and feeding place—it would always end up with breakfast (portrayed on the left side of the Figure 7.2).

To demonstrate that was what the bird was doing, an investigator simply blocked the view of the ersatz sun with a small screen and used a mirror to reflect its image into the cage from another direction, creating the impression that the sun had instantaneously moved to a new location. The bird, on release into the arena, not catching on to the ruse, now held the same 90° angle with the sun's refulgent image in the mirror, and this led it to an empty feeder (see the right half of Figure 7.2).

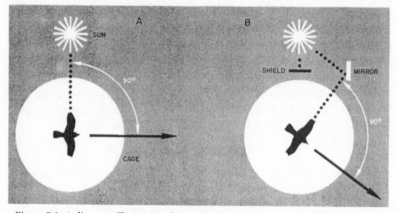

*Figure 7.2. A diagram illustrating the results of a simple, but ingenious, experiment done to demonstrate that birds use the sun to orient themselves in space.*

That was a clever and easy experiment to do, and the results were sound. Birds use the sun to orient themselves in a particular compass direction. Unsurprisingly, the behavior has been named sun-compass orientation. Now let's take the idea outdoors and apply it to a bird migrating south in the fall. As we learned above, for a bird to travel south at 9 A.M. it will have to fly at an angle of 45° to the right of the sun. To continue on its southerly journey at 3 P.M., the same bird will have to assume and angle with the sun that is 45° to the left of it (see the panels a and b in Figure 7.3). With the wisdom of a wise old owl, the bird does just that. As you can see, there is much more to flying south than just learning one angle to assume with the sun. In short, it is difficult to apply the sun-compass idea to a natural situation because the sun is not a stationary benchmark—it is moving across the sky from east to west at 15° per hour. Without dragging you through the excruciating work that led investigators to the answer, I'll just come right out and tell you that the bird uses its living clock to compensate for the steady movement of the sun across the heavens. That simple statement should set your minds to wonder. Aren't you tempted to speculate that even young birds, which may have pecked their way out through an egg shell a few months earlier, have quickly mastered geometry, astronomy, and astrophysics and are thus fully cognizant of direction and rate of passage of the sun across the heavens each day? We may have to redefine the conventional meaning of birdbrain.

Just how birds find their way to a specific geographic location (some have the accuracy of an intercontinental ballistic missile) during annual migrations is very complicated, but certainly sun-compass orientation

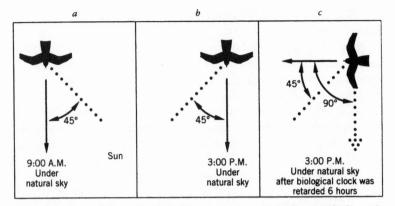

*Figure 7.3. The need for the living clock to use sun compass in direction finding. The first two panels show the requirement of assuming different angles with the sun in order to maintain a constant southern compass bearing as the sun moves across the sky during the day. The living clock provides the time-of-day information needed to do this. When the clock is reset artificially in the laboratory and the bird released, it uses the bogus time of day indicated by its clock and flies off in the wrong direction (last panel).*

plays a role. A question then arises: Is the clock that is used in this orientation the same one used to control their activity rhythms? Here is the story. A bird that was using the real sun to fly south along its fall migration route was brought indoors and subjected to artificial days in which sunrise and sunset were moved six hours later, meaning that 9 A.M. in the bird room, now occurred at 3 P.M. Mother Nature's time. As soon as the bird's activity rhythm had adjusted to this new schedule (its clock had been reset by 6 hours) it was taken out into the real world at 3 P.M. and its direction of flight tested. Now go to panel c of Figure 7.3 and see that when the bird was released it chose a due west course (rather than the southerly migratory direction) because it interpreted the sun as being at its 9 A.M. position in the sky. Thus it is clear that the living clock, the same one that times bird activity rhythms, also plays at least this much of a role in bird migration.

One of the major problems in studying migration is that the subjects quickly fly away and you lose them. One could, and some have, followed migrating birds in planes, but that can be expensive (arctic terns migrate 22,000 miles a year!), and it is very easy to lose sight of one's quarry from a plane . . . there's a lot of sky up there. One way to solve the problem is to use flightless birds, living in open country where they can be easily followed, such as penguins—which provide the additional benefit of walking very slowly (ostriches were considered but they can run 42 miles per hour and are nasty when you catch up to one).

Penguins were captured in Antarctica at a seaside rookery and transported many miles inland by snowmobile and released. The releasee does one of two things: it takes a nap (in which case you have to sit in the bitter cold and wind and wait for the siesta to end), or it stands for a few minutes as if getting its bearings, and then waddles off. No flashing wings here and a bird quickly vanishing in the distance; it may take a penguin an hour before it disappears over the snow-covered horizon. And then, if there is fresh fallen snow on the ground you can follow the trail without the bird ever knowing you are trudging along behind it on snowshoes. On overcast days (which are all too common in Antarctica) the penguin is disoriented and wanders aimlessly. But when the sun is visible and circling overhead (during austral summer it never sets), all translocated penguins go north. Always due north, even though that direction seldom leads directly back to the bird's relatives in the rookery (thus this response cannot be called homing). But the orientation is adaptive, because these birds literally live at the end of the earth, so going north always takes them to the sea that holds their only source of food. When they do reach the water and dive in, they then began to navigate (on land they only compass orient using the sun) and literally fly home under water. One penguin was carried 1,200 miles inland but still made it back to its own nesting site . . . however, the trip took ten months!

Now envision in your mind's eye a standard globe of the earth. Vertical meridian lines, 15° apart, run from top to bottom, dividing the globe into 24 slices. It takes an hour for each slice to pass under the sun as the earth turns on its axis. At the equator, these meridian lines are most widely separated: There the distance between each of them is 1,038 earth miles. But the width between them gradually lessens as they extend southward and drops to nothing when they all fuse into a single point at the south pole. On the equator one must travel over 6,000 miles to cross six time zones, but in Antarctica, while the Russian outpost at Mirny is six time zones away from the Cape Crozier on the Ross Ice Shelf, the distance is less than a third of that—easy transport distance for a load of penguins. Thus, penguins were quickly carried from Mirny to Cape Crozier and released. The birds' clocks were set to Mirny time, so when released, after a quick look at the sun they set off in a westerly, rather than a northerly, direction. But if they were kept in a pen at Cape Crozier for three weeks (while their clocks reset to local time as this experiment was done at a season when there were day/night cycles) and then released, they went north, the patent direction of deported penguins.

## STUDIES IN BEELINES

Humans have a certain ambivalence about bees: We love the honey but fear the sting. Fortunately, there is a handful of biologists who are undaunted by the latter. And one of these intrepid souls, Karl von Frisch, won a Nobel Prize in 1973 for his work on bees' behavior.

The story I am about to tell began with a Swiss naturalist, August Forel, who on nice summer mornings chose to eat his breakfast on his open porch. He never ate alone; he was joined by neighborhood honeybees that came to gather his marmalade. When Dr. Forel ran out of marmalade he noticed that the bees continued to come at breakfast time, only, for several days, but then gave up. But when he finally got to the store to replenish his supply the ritual slowly began again. He mentioned his observation to a contemporary who retorted that he had noticed that bees only came to his buckwheat field during the early-morning hours when the buckwheat flowers were open and secreting nectar.

The first experiment to be designed around these observations was done in 1929. A table was placed in a field with a dish of sugar water on it; a lionhearted, note-taking investigator, Ines Beling, then a student of von Frisch, sat nearby. Eventually bees bumbled on to the treat and while they lapped up the sugar water, she painted different colored spots and patterns on their backs. This early version of bar coding was used to identify when and how frequently a bee returned to the feeder. The sugar water was offered only between 4 P.M. and 6 P.M. each day for about seven days, and each of these days the number of bees visiting the free-lunch counter increased (there must be a message here for socialist governments). Then the welfare was abruptly stopped. Hooked on a freebee system that didn't even require shuffling food stamps the bees continued to come for a while. Not finding the expected prize, the bees landed on, and tromped all over, Beling's body searching for hidden sugar. Being made of the right stuff (meaning not nectar), she did not flinch or flee, and her notes don't even mention the number of stings she received (but she did record the bar-code numbers of her potential assailants).

Nor did she give in to their demands for six straight days, during which the bees returned mainly between 4 P.M. and 6 P.M.; then they stopped visiting, being forced to return to visiting beautiful, sweet-smelling, nectar-laden flowers. We now know that it was the bees' clocks that were directing their tenacious return each day at the learned time. Beling continued her studies and found that she could train the bees to forage at several different times during the same day as long as she spaced the test times a few hours apart. She even taught them to go to different collecting sites at different times of the day. She tried to teach them to arrive at times other than 24-hour intervals. Nineteen and forty-eight hour durations were tested, but the bees could not handle either interval. Those exposed to 19 hours, just came at all times of day, and those subjected to the longer interval came every 24 hours . . . as if they had learned to divide by two. These observations show the deep-seated nature of the 24-hour interval in the timing of the honeybee.

Several studies done over the years—like the one that suggested that earth tides may influence cave cricket activity rhythms (Chapter 6)— spurred some investigators to hypothesize that a subtle, periodic geophysical force that could not be, or was not, eliminated from otherwise laboratory-constant conditions was providing the needed timing information to the organisms incarcerated there. The honey bee foraging rhythm was used as a means of answering this question. Two identical bee rooms were built, one in Paris and the other in New York.

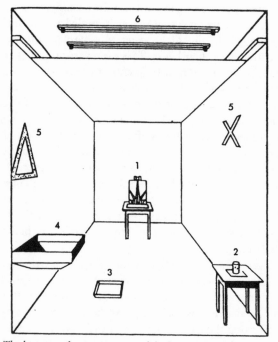

*Figure 7.4. The layout and accouterments of the bee room. 1—hive; 2—sugar-water table; 3—BP pan; 4—drinking water; 5—orientation markers on the walls; 6— indirect lighting. Light and temperature were held constant in these rooms.*

A colony of bees was trained in the Paris room to visit the sugar-water table between 8 A.M. and 10 A.M., and after they had learned the routine they were flown overnight to New York, five time zones away. The idea was to get 40 bees across the Atlantic and into the identical room as quickly as possible, but permission was required by the U.S. Department of Agriculture, the Customs and Immigration Services, the New York Police Department, and the American Museum of Natural History (this experiment was done in 1957; just think of the additional layers of bureaucrats that would have to be penetrated today). Thus the 8-hour trip took 20 hours. The bees, that had not been permitted to see a day/night cycle in New York, were released into the bee room, and everyone sat down to wait and see what time they would visit the sugar water. If they did so between 8 A.M. and 10 A.M. Paris time, it would mean that the bees thought they were still in Paris. But if they searched at 8 A.M. to 10 A.M. New York time (which was 1 P.M. to 3 P.M. in France) then they would be responding to some exotic, time-giving geophysical force that had penetrated into the bee room in New York. They used Paris time to venture out of their hive, strengthening the idea that the clock was like our windup version whose escapement mechanism generates its own time, rather than like a sundial that gets its timing information from the passage of the sun overhead. The bees were clearly not getting their timing information from some periodic, geophysical force able to penetrate the laboratory walls.

Although the rest of the book could be devoted to the behavior of bees, I'll describe just one other use of their clock. I mentioned above that Ines Beling was able to train her bees to visit a feeding station at several different times in the same day. Bees, like birds, find their way around using sun-compass orientation, so her bees had to have used a different orientation angle for each of these differently timed visits. They must have a clock able to compensate for the movement of the sun across the sky. This was proven by simply waiting until trained bees arrived at a feeding station, capturing them, holding them in a dark box for several hours, and then releasing them. Remember that bees first fly to a station using the position of the sun in the sky. After collecting their booty in a few minute's time they then do a few calculations and determine the angle needed to get back to the hive. But when trapped in a box for three hours the sun has moved 45 degrees across the sky; if the bees used the return angle they had previously calculated, they would fly off in the wrong direction when released and not return to the hive. But they do not, they go directly home (taking a bee line) because their clocks have compensated for the movement of the sun while they were boxed up.

Here is a variation on that theme. As you may already know, when a honeybee finds a good source of nectar she tells her hive mates how to find it by doing a dance on the hive's honeycomb. The orientation of the dance she assumes on the comb signals the direction of the last source of nectar she found. Sisters feel her dance with their antennae (it's too dark in the hive to see) and then fly off directly to the correct flower patch. If it is a rich source that is available during an entire morning the bees take full advantage of its availability. Loaded bees return later in the day and dance, constantly changing dance orientation to compensate for the apparent movement of the sun across the sky. A few bees, after finding a nectar source, return to the hive and just continue to dance day and night. They are called marathon dancers. Because the dance signals give the direction of the last nectar source they knew about, and earth constantly spins under the sun, the direction signaled by the bee's dance rotates on the comb once every 24 hours. The end impression is just like having a living wall clock in the hive!

## LAND CRABS AND PANTHERS

At one time in my career I temporarily moved inland from the shoreline to study the rhythms of terrestrial crabs. Working at the well-appointed Lerner Marine Laboratory on Bimini (an island in the Bahamas) I used the great edible land crab belonging to the genus *Cardisoma*. It is a mean monster (among crabs) with dangerous (but tasty) pincers. In nature these crabs remain in their deep burrows during the daytime, but emerge to 4F at dusk; their rhythm is a daily, not tidal, one. One of the specimens I had collected happened to be a super crab: In the constancy of the lab it became active each day at 6 P.M. I bragged about the precision of its timing (I am chagrined to reveal that sometimes when in a schoolmarmish blue funk over the deteriorating American education system, I am more impressed by the performance of a few of my crabs than some of my students) and was immediately challenged by colleagues to risk a bet on my star. The bet was of the sucker variety: If super crab became active at 6 P.M., plus or minus five minutes, they would buy drinks for me that night at the wonderful, postage-stamp-sized End of the World Saloon on this out-of-the-way island. Otherwise, I bought theirs. The contest lasted for ten nights, and super crab never failed me once.

One of the high points of those evenings was the presence of Adam Clayton Powell, a New York Representative to the U.S. Congress, who, having his own retreat on Bimini, used the End of the World as his watering place. He had a reputation as a scalawag politician; he was in and out of trouble, and he was loved by some and not loved by other New Yorkers. Here, out of the political limelight, he became his other self: a delightful, intelligent, entertaining, colorful addition and friend to our group. He also recognized a good thing, so he bet on super crab.

Having collected an enormous amount of laboratory data on the edible land crab, I decided that before publishing it I should carry out a quick study of the crab's behavior in its natural habitat to see if our lab data were representative. Late one afternoon I took a small boat over to neighboring South Bimini Island, dragged it ashore, and set off into the forest inhabited not only by hundreds of land crabs, but also, according to the natives, by at least one black panther. Both crabs and panther are active at night. Before the sun set I had spread a smooth layer of flour around the entrances of many crab burrows (needless to say I used "all-purpose" flour), and then sat down on the ground for the rest of the night. At one-hour intervals I turned on a red light and checked the flour around each hole for crab footprints. Finding spoor indicated a burrow's resident had emerged and was out on the town. I smoothed over the tracks in the flour and sat back down in the pitch darkness for another hour.

After about four hours the freshness of this enterprise had pretty much staled. Sitting in the dark was no fun, and I was getting bored and tired; I let my mind drift to thoughts of the panther that was probably wandering about out there somewhere. Deep in a rain forest in pitch darkness one feels, and is, awfully alone. What would I do if I turned on my red light and found a panther print in the flour! While ruminating that adventure, I heard the first sounds of what had to be a large animal nearby scavenging for an easy meal. Stealth was impossible because each step it took rattled the dry fallen palmetto fronds strewn everywhere. The steps were getting closer. I weighed the options of the most fundamental law of jungle survival: fight or flight. I was cocksure I couldn't outrun a panther, and all I had to defend myself was a wee flashlight. Solution: As the animal pounced I would shove my arm down its throat causing the cat to choke to death. I would probably lose my arm, but only Dr. Heimlich could save the cat. Now with a viable game plan I was less terror stricken; I tore the red cellophane off the flashlight and sent its beam toward the killer. But I digress; Sorry.

My adrenaline rush declined slowly, so I had no trouble staying awake (and listening) the rest of the night. By dawn all of the crabs had traipsed back home, leaving their scuff contrails through the flour as they reentered their burrows. Ignoring any panthers and me, they duplicated the same activity pattern out there as they had exhibited in the lab.

Oh, in case you're wondering, when I turned on my light at that disconcerting moment mentioned above, I found no panther. Instead, shuffling through the brittle floor litter were large land-dwelling hermit crabs (*Coenobita clyeatus*) each housed in a heavy turret shell it had appropriated from the snail that made it. Every now and then one of these top-heavy crabs would topple off a fallen log it was attempting to navigate and thump to the ground, making a foot-drop sound. I had allowed myself to be dragooned by an animal so insecure that it spends its cowardly life hiding inside a snail shell.

## SMALL MAMMALS

Some of the most rewarding subjects used in the study of biological rhythms have been mice, both the kind bred for the laboratory and the very handsome, white-footed deer mouse. The latter lives in the wild but is also willing to share our homes, so you have probably noticed the fine texture and beautiful coloration of its coat when you remove a squashed, pop-eyed one from a trap. The popularity behind the laboratory use of wild mice is that they are easy to maintain, they have very accurate clocks, and it is quite easy to record their activity automatically so that one need not often interrupt otherwise constant conditions. Also, we can let the wild ones go after they perform.

A typical activity recorder consists simply of a cage with a running wheel attached. The cage is provided with sufficient food and water to last the duration of a study, and it is self-cleaning in that the floor is a grating that urine and feces drop through into a litter box. While in captivity mice spend their time sleeping and eating in the cage, or running in the exercise wheel. They have a strong penchant for running in place in these treadmills-to-oblivion and are known to run as many as nine miles in a single night! The apparatus and a set of data are shown in Figure 7.5. The activity shown in Figure 7.5 was nicely rhythmic (note the precision of the onsets of activity); it persisted for 25 days and had a period slightly less than 24 hours long.

Physical activity makes an individual's heart beat faster. Thus, accompanying a rodent's activity rhythm is a changing heart-beat rhythm. Certainly the peak of the heart-beat rhythm is largely a consequence of strenuous exercise, but the pulsation cadence is also under some clock control. In demonstrating the latter's involvement the heart of a hamster was removed and placed in a solution that kept it alive and beating for as long as 38 hours. The contraction rhythm persisted. You may be thinking that this is not so strange because the pacemaker (that some of us have an artificial replacement for) within the heart is still functioning. That is true, but it is not the whole story by any means, as the following experiment demonstrates. This endeavor required an extreme manipulation: Trypsin, the digestive enzyme found in our stomachs, was used to separate the individual cells of a hamster heart (the cells were not digested by the treatment, just the material holding them together). Thus chemically dissected, the heart cells were then submerged in a special culture

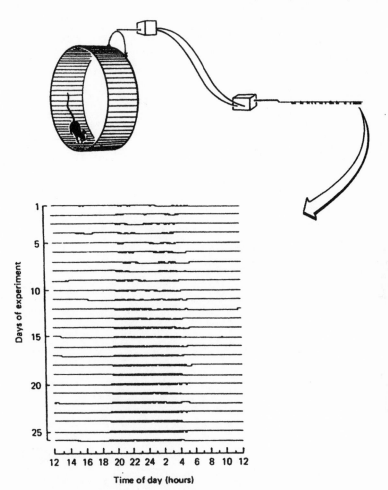

*Figure 7.5. How activity records can be made. Each time the running wheel turns once, the event is recorded as a pen scratch on the chart. When a mouse is very active the scratches fuse into a solid band. Consecutive days of activity records are plotted one beneath the other.*

medium in which they survived quite well. After about three to five days the cells had grown together into little clumps, and each clump, beat as a unit. When the changing rate at which the clumps beat was measured it was found that the daily rhythm had survived intact! Combining this finding with those I described in single-celled algae and protozoans in Chapter 1, there is no reason not to think that each heart cell has its own clock! Certain aspects of biology can be stranger than fiction.

C H A P T E R   E I G H T

# A Few Plant Clocks

### Do Plants ʟᴇᴇᴘ?

P robably the first written record of a biological rhythm comes to us from the pages of a diary kept in 350 B.C. by a centurion in the corps of Alexander the Great. A man of many talents, the officer also amused himself by studying plants. It was he who first noticed that some plants performed what are today called—just as erroneously as in the past—sleep movements. At night the blades of the leaves are dropped to the sides of the stem (as if asleep), but at the next dawn they are lifted high again, as if in a pagan gesture to the rising sun. They remain in this worship position all day and then doze off again at sunset.

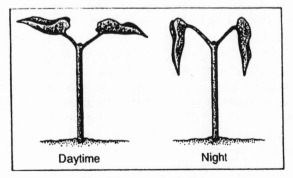

*Sleep Movements in the Bean.*

For the next 2,079 years naturalists and botanists simply sat around marveling at the miraculous ability of some plants being able to move on their own, but sought no answers as to how. It wasn't until 1729 that an astronomer took up the challenge and brought plants indoors. To his great surprise and discombobulation, their sleep-movement rhythms continued in the absence of day/night alternations in his basement laboratory. It must have taken great courage for him to publish such an observation. Even though this was the first evidence suggesting the existence of a living clock, his work went unrecognized for another 100 years, probably because he was a sky gazer rather than a member of the botanical brotherhood.

When the botanical experimentalists did begin to act, a great deal of interesting and important information blossomed. Even Charles Darwin jumped on this bandwagon; he was an insomniac and thus easily made many nighttime, as well daytime, observations on plant movement. Darwin recorded what he saw for posterity in his 1881 book, *The Power of Movement in Plants*. Eventually it was noticed that the period of the rhythm usually lengthened a bit in constant conditions, and this led to the understanding that it was the day/night alteration that forced the rhythm to a 24-hour length when plants were in their natural setting. Using bean plants it was discovered that there were many clocks in a single plant, and that they did not all necessarily have to be set to the same time. In demonstrating this a plant was kept in normal day/night conditions but one leaf was shielded from it and instead offered a reversed night/day regimen for several days. The lone leaf thus went up while the others went down, and vice versa. Then when the plant was put into constant conditions the reversal continued for several days signifying that each leaf had its own clock, rather than the show being run by a single master clock. But after a while the odd leaf either drifted back into phase with the others, or simply crumpled, giving up its temporal ghost.

Just where in the leaf the clock resided has also been found. The movements are brought about by two packages of cells, each called a pulvinus. They resides on the upper and lower sides of the hinge junction between the petiole (the leaf stem) and the spade-shaped blade. The pulvini take turns, one swelling with water, while the other is deflating. When the lower one swells it acts like a hydraulic ram and lifts the blade up. When the one on top swells and the one on the bottom drains, the blade drops to the sleep position. It is the plant's clocks that give the signal when to swell and shrink by altering the permeability of the pulvini membranes to potassium and other salts. When the salts pour into pulvinal cells it causes the osmotic uptake of water that brings about swelling.

Some early investigators could not accept the existence of a living clock within plants. They felt that plants must somehow be getting the necessary timing information from the environment. The only way to explain this was to propose that some periodic, time-giving geophysical force must be able to penetrate into standard laboratory constant conditions, meaning it had to pass right through the roofs and walls of buildings, and whatever container a plant might be kept in. One such believer tested this idea by taking his bean plants into a 487-foot-deep abandoned salt mine in Germany to see what would happen to their sleep-movement rhythm. The result was unambiguous: The rhythm stopped, then restarted again when the plants were returned to the surface. Was that proof that shielding one's test organism with a tenth of a mile of dirt and rock caused arrhythmicity? The finding stood as a curiosity, unexamined by others for 33 years.

But then a student of mine decided that he must make an attempt to repeat that old work, so we made a battery-driven recorder that would measure plant sleep movements in the bottom of a mine shaft. Ready to try a repeat, he now had to meet the requirement of getting to the German salt mine. His solution was near perfect: He married a beautiful, very bright woman, and they honeymooned in Germany. When they returned, the new bride happened by the lab; after saying she looked radiant and all those perfunctory things one is expected to say at times like this, I asked her how she liked Germany. Her answer was, "Wonderful, but thanks to your guys' interest in those damn biological rhythms, the connubial bliss was often interrupted when my husband/pocket-protected scientist repeatedly dragged me down in that salt mine!" A new student in the lab who had not yet been briefed on the salt-mine project, overheard her comment and ejaculated, "Wow, kinky!"

It's best to leave the now-blushing bride and tell you what the beans did at the bottom of the cave . . . nothing. There was not the slightest sign of a rhythm in the bowels of the cavern. Thus, the early work was confirmed, but the interpretation was entirely different. It was not that a mysterious, periodic, time-giving geophysical force had been screened out by the thick layer of dirt and rock over the deep mine experimental chamber. It was the salty air. When the early study had been done the role of salt movement into and out of the pulvini was unknown. Therefore, the facts that the experiment was carried out in a damp salt mine, and that the plants were enveloped in humid salt-laden air were ignored. Armed with the latest information on leaf movements our underground newlyweds looked for and found that the salty air caused the cessation of all leaf movements. Case closed.

A leaf torn off a plant can be kept alive for a while in water, so, if each leaf has its own clock, an isolated leaf should continue to undergo a blade-flexing rhythm. That is true, and the evidence is shown in Figure 8.1.

The petiole of a plucked leaf was simply inserted through a hole in a rubber stopper into a nutrient solution in a test tube, and the preparation was put into constant conditions. The isolated leaf's rhythm persisted for 28 days! Even more drastic treatment does not discourage a leaf's rhythmicity. All of the blade tissue can be cut off, leaving only the midvein, and that preparation continues to be rhythmic also. The mutilation was carried even one step further without destroying the rhythm. Pulvinal cells were dissected out of the plant and then carefully separated from each other and kept alive in a special culture medium. The swelling and shrinking rhythm in each cell continued for as long as the pulvinal cells remained alive (about nine days), meaning that each has its own clock.

In some plants in addition to the pair of pulvini discussed above, there is another hinge at the insertion point of the petiole and plant stem. Both pairs function so that there is a dual action to an overall leaf's motion: the whole leaf lowers at the stem, and the blade also folds down. And when kept under constant illumination, if the light intensity is adjusted just so,

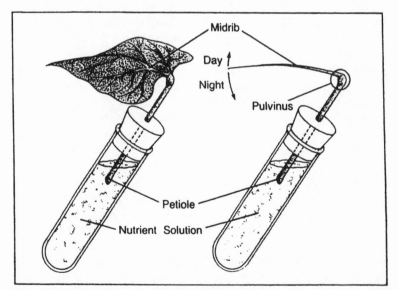

Figure 8.1. An experiment that demonstrated the persistence of the sleep-movement rhythm of a single leaf (left) and of a leaf in which the blade has been shaved off leaving only the midrib attached to the petiole (right).

the stem/petiole sleep-movement rhythm assumes a circadian period that is different than the blade/petiole rhythm: One is 28 hours and the other 30.3 hours long. Thus, two living clocks in the same plant can be made to run at different rates in the laboratory.

Remember that the previous chapter contained a discussion about being unable to train bees to come to a feeding station at intervals other than 24 hours. This type of experiment has been tried with the sleep-movement rhythm also. Plants were subjected to artificial day lengths, such as 6 hours of light alternating with 6 hours of darkness, or 24 hours of light flip-flopping with the same length of darkness. Even though the plants were exposed to many repetitions of these abnormal cycles, and sometimes appeared to be entrained to them, when they were then moved into constant conditions, the period the plants displayed was always close to 24 hours.

*Bryophyllum,* a common house plant, undergoes a rhythm in the rate at which it fixes incorporates into organic molecules) carbon dioxide ($CO_2$) using a method other than photosynthesis. The details are complicated so please just take my word for it since that is not my purpose for introducing this plant. My reason is to point out that the $CO_2$ is fixed rhythmically, and this rhythm will persist in an isolated leaf, even after the upper and lower epidermal tissues have been removed. The cells remaining after the stripping are mainly just photosynthetic machines, yet each apparently come provided with a clock to guide this special kind of $CO_2$ fixation.

It is really more common for flower petals to open and close than for leaves to go up and down. Morning glories are so named because they open their petals in the morning (to attract insects for pollination purposes; not for humans' esthetic enjoyment), but in fact some morning-glory varieties, like Moon Vine, open (in a matter of seconds) in the evening. Those misnomers notwithstanding, in the seventeenth century it was somewhat common in Europe to have clock gardens. These plots had groupings of flowers that each opened at fairly specific times of day. If it was sunny day, looking out a second-story window down into the garden to see which flowers were open, could insure one of being sufficiently tardy to a party to appear appropriately sophisticated.

The petal-movement rhythms are also under the control of a living clock. The night-blooming jessamine, a tropical shrub, opens its flowers at night and emits a powerful fragrance that attracts nocturnal-flying insects, who, unbeknownst to them, transfer pollen between flowers while collecting the sought-after nectar. Cut flowers from this plant brought into the house continue their rhythm (therefore screens on the windows are a must). Because the odor is not noticed until the flowers open, it was logical to assume that the petal movements allow the fragrance to escape. As is too often the case, that interpretation is wrong. Plucking petals from the flowers and floating them on water gives the correct answer. The substance responsible for the pungent odor is released from the petal tip, whether the flower is open or closed, and the rhythm continues in an isolated petal until the petal dies. The jessamine's clocks just happens to control petal movements and fragrance emission simultaneously.

Not all botanical work on rhythms focused only on the movements of leaves and petals. Before the turn of the twentieth century it was known that if a sunflower stem was cut off just above ground level the sap continued to be pushed up by root pressure and expelled from the decapitation wound. When collections of the exudate were made at constant intervals through the day and measured and compared, it was found that the amount varied rhythmically between night and day. It was also known that plant growth was faster at night than in the daytime. Both of these rhythms persisted in constant conditions.

## A PLANT/FIREFLY

Now, as a result of the last 40 years of intensive experimentation, it is known that just about every basic aspect of plant life is controlled by living clocks. The study of some of these rhythms is very labor intensive, so much so that the effort needed to define a rhythm leaves little time to examine what the clock is doing at any moment, and that is the ultimate goal of chronobiology.

A tiny weed that has become very useful to cell biologists is called *Arabidopsis*. It is rather easily cultured in Petri dishes and much of its genetics are now known, including the fact that a few of its genes are responsible for the rhythms it expresses. Some of its rhythms are difficult and time consuming to study: for example, if wanting to learn when the daily maximum in output of some substance occurs, test plants have to be ground up at regular intervals through the day and night, and substances extracted and analyzed. This tedium could be avoided if there was a way to know in advance what time the clock was signaling. The investigator would then be able to do a single extraction and be sure of collecting a maximum amount of the desired product. With that information it would be possible to assay thousands of plants in a short time. That can now be done, but first some brief necessary background.

It is genes that initiate most everything that goes on in cells. They turn on and off as needed. Thus, when, say, insulin is needed, the gene, or genes, required to make it available turn on. Bioluminescence in fireflies is produced by a chemical reaction that is gene controlled: for the tail lights of the flies to light up, a substance called luciferin is combined with oxygen in the presence of the enzyme luciferase. The key to flashing is the presence of luciferase, and this activity is under the control of a specific gene.

It has become quite newsworthy of late that biologists now routinely remove genes from one organism and put them into another, often an entirely different species. In the case at hand cell biologists Andrew Milnar and Steve Kay moved the firefly luciferase gene to *Arabidopsis*, attaching it close to a clock gene. Now, when the clock gene is turned on so is the luciferase gene, and the plant would glow like a firefly if it only had luciferin . . . but it does not. The trick then is to supply it, and this is done by simply spraying it on the plant. Thus at any time the investigators wish to know whether the *Arabidopsis* clock is on or off, all they have to do is squirt a puff of firefly luciferin into a Petri dish filled with the little plants and see if they shine.

## PLANT PHOTOPERIODISM

The following account serves as well for birds and insects as it does for plants, but in this chapter we focus on the latter. As we all know, most plant species do not flower all year long; instead they choose a season. Bloodroot, trillium, crocuses, and others are spring flowers; while asters, ragweed, and the other hay-fever horribles flower in the fall. Iris and columbine flower in the summer. Skipping a long, tortuous, and very interesting Sherlockian investigation that finally revealed how this seasonality is controlled, I'll just give the one-word answer: photoperiodism. The relative number of hours of daylight versus darkness during the interval of a day controls flowering in many plants, reproduction in birds, and reproduction and certain aspects of development in insects and mammals. It is the short days (when hours of daylight are reduced) of spring and fall that stimulate trillium and ragweed to flower, and the long hours of daylight in the summer that cause iris and columbine to set flowers. Thus these plants and others are grouped into short-day plants and long-day plants (botanical nomenclature is really easy).

I'll give just one example, in this case a pioneering one, of the kind of experiment that proved this. The eternally popular (but illegal) plant, *Cannabis,* only flowers when the hours of daylight are relatively short. Entrepreneur marijuana pushers and home-gardener users, contrary to government desires, would like a year-round supply of flowers. The answer as to how to do this was present back in 1912—a man named Tournois learned that if he covered his *marijuana* plants (meaning more than just being in flower pots) during the daytime with boxes, so they received only six hours of sunlight, they would flower during the long days of summer. We do not know if he benefited from this discovery because he was killed early in World War I, and his work went mostly unnoticed by dopers and botanists.

The fact was rediscovered in 1920, and by the 1960s many flowering plants had been studied under a variety of photoperiods and had thus been pigeonholed as short- and long-day varieties, plus a day-neutral category that will flower without regard to the ambient photoperiod (to be discussed briefly later). This information was so widely known that high school biology classes did experiments so foolproof that they actually worked: the instructor was sure that if his students exposed, say, a short-day trillium to artificially created short days that it could be made to flower at any time of the year. But then, the experiment was modified by a spoilsport who turned on a dim light for a short time during the middle of the long dark period, (as shown below). That simple change inhibited the short-day plants from flowering. To everyone's surprise, it was the length of the night interval that the plants were measuring—not the hours of daylight.

of course, not flower. But if the long night is interrupted with light briefly at its midpoint, then it will flower. Therefore, like short-day plants, long-day plants also measure the interval of nighttime. It turns out after years of talking about photoperiodism, it was not the light interval that the plants were using: it was the number of hours of uninterrupted darkness. The flowering response should have been called scotoperiodism (scoto = darkness)! And short-day plants should have been called long-night plants, and long-day plants changed to short-night ones. But it was too late, photoperiodism and the rest, though now known to be technically incorrect, was as ingrained as were other malapropisms, like a fiddler crabs that can't play a note, a cave cricket that isn't a cricket, a morning glory that blooms in the late afternoon, a Department of Defense that makes war, wars that are called police actions, and all the rest of the nuttiness and spin we are subjected to daily.

The fact that dim nighttime illumination can be very important to the flowering of some plants may be the significance of the sleep movements they perform. It is the leaves that are the receptors of light, and if they droop at night less light can impinge on them during the nights around full moon.

I would hope you readers are all thinking, "How do plants measure the length of the night interval?" If you bet they're using their living clocks, you are correct. Here are the kind of experiments performed to give us this conclusion. The explanation will take some extra words of introduction before an understanding can be complete.

Employ Figure 8.2 for this discussion. Using the old, ingrained terminology, short-day plants were exposed to days consisting of 6 hours of

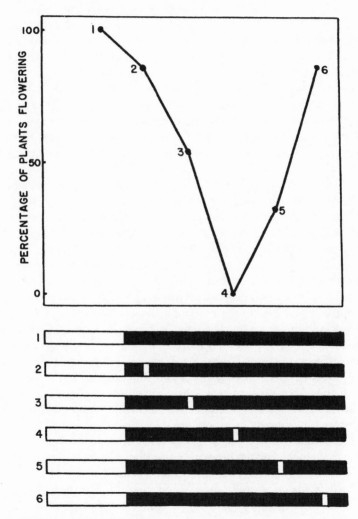

Figure 8.2. The experimental design used to demonstrate the changing sensitivity (top half of the figure) of a particular species of plant to the timing of brief light breaks at various times throughout the night (bottom half of the figure).

light alternating with 18 hours of darkness (the condition seen in the horizontal bar labelled one just under the box in the figure). The subjects, being short-day plants, all flowered. This fact is recorded by the point labelled one, at the 100 percent flowering level, on the curve above. Then, all of these plants were discarded, and the a new group was tested under the same photoperiod, but with one variation, the dark interlude was interrupted with a 15-minute light break about two hours after the time the light was turned off (see bar two). This interruption was sufficient to inhibit 15 percent of the plants from flowering (see the level of point 2 on the curve above). That group of plants was then discarded, and batch after batch of new ones were tested at the dark interruption times indicated by bars 3, 4, 5, and 6—the points above each indicate the degree of flowering disruption. As expected by the results reported two paragraphs above, all flowering was inhibited in the photoperiod indicated by bar 4. Thus, the degree of flowering reduction depends on the time at which the dark period was interrupted by a light pulse. Another way of saying the same thing is that for short-day plants there is a regular changing sensitivity (such as, the degree of inhibition) to identical light breaks during the night.

This type of experiment has been repeated many times, using both short- and long-day plants, and more importantly, in ways that give a better answer as to what is going on. Here is the type of examination that elucidated the role of the clock in photoperiodism. A 70-hour "day" consisting of 10 hours of light alternating with 60 hours of darkness was created in the lab. Kalanchoë, a common short-day house plant, will flower when maintained in such an artificially long, short day. Then experiment after experiment was carried out patterned the same way as the approach described in the paragraph above—the dark period was systematically broken by short light pulses. Literally hundreds of plants were tested and then discarded. As above, flowering occurred some times, but not at others (Figure 8.3). Note that flower inhibition occurred at roughly 24-hour intervals . . . suppression was rhythmic! Kalanchoë's clock drives a rhythm in sensitivity to light: When light pulses are given at the peaks of this rhythm, flowering is inhibited. A graphic representation of this rhythm is portrayed by the bold sinusoidal line in Figure 8.3.

The same light-sensitivity rhythm exists in long-day plants also, only the peaks in this rhythm promote, rather than prevent, flowering—that is the difference between long- and short-day plants.

There is a great deal more evidence that the clock functions as described, but I will not go into that. Instead just study Figure 8.4, which is a diagrammatic representation of the light-sensitivity rhythm (the bold line shaped into a flat-topped pyramid) and its clout in action. On the

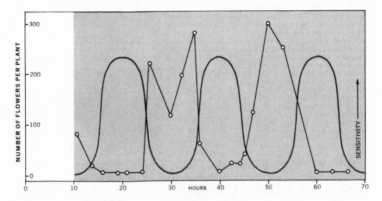

*Figure 8.3. The differential induction of flowering in the common houseplant Kalanchoë when exposed to short days, consisting of 10 hours of light alternating with 60 hours of darkness (the long dark interval was systematically interrupted with brief light breaks). Note the suppression of flowering at 24-hour intervals (the curve based on open-circle points). Superimposed over this curve is a smooth curve representing a clock-driven cycle governing the sensitivity of the plant to nighttime light interruptions.*

left-hand side of this figure we see the sensitive phase of the rhythm in both short-day and long-day plants occurring during the nighttimes of spring and fall; thus short-day plants flower but long-day plants do not because no light falls on the stimulative part of their rhythm. On the right-hand side the nights of summer are not long enough to enshroud the entire peak of the sensitivity rhythm, so long-day plants flower while short-day ones do not because the tails of their rhythm's peak are illuminated, stifling flower production. The figure is an oversimplification, but it should convey the basic idea of how the clock and changing photoperiods, together, cause a seasonality in flowering.

In case you are wondering, not all plants regulate their flowering to changing day lengths. For example, the so-called day-neutrals, like tomatoes and roses, will flower over a wide range of photoperiods as long as other environmental conditions are propitious. Still other plants (and especially animals) have an annual clock that guides their reproduction schedule. However, not a great number of investigations into the existence of annual rhythms have been undertaken. There are several reasons for this. First, to make sure a process is really rhythmic requires seeing a cycle repeat several times, and in an academic environment in which publish or perish is a pragmatic guideline dangerous to ignore, few people (and none of the yet untenured) are willing to devote a minimum of three years to a

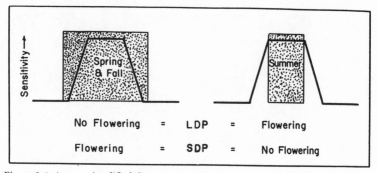

Figure 8.4. An oversimplified diagram created to compare the role of the light-sensitivity rhythm causing flowering seasonality in long-day (LDP) and short-day (SDP) plants.

single study. Secondly, it is very difficult to hold laboratory conditions constant for several years (severe storms cause electrical power outages; grants run out; and a hundred other unexpected problems). Fortunately, some investigators are willing to accept such risks, and a few annual rhythms are known. I will describe just one: an annual rhythm in seed germination.

Digitalis and chrysanthemum seeds were stored in constant darkness and unvarying temperature for several years. Then at regular intervals a few seeds were removed and subjected to ideal, identical, germination conditions. The number that sprouted was recorded. The germination percentages were greatest in the early spring and least in the summer. Clearly there are living clocks in plants that measure off approximate yearly lengths.

# NINE

# DENOUEMENT: THE LIVING CLOCK

Over the years biologists exploring the terra incognita of organismic timekeeping have had some successes; I can thus introduce much of what is known about the living clock's escapement. First, a bit of historical perspective.

I first got interested in rhythms when, as a student, I stumbled upon my first example. As a beginning animal physiology student (the hot field back in those days) I had learned all about homeostasis, the name for the fact that warm-blooded animals are able to maintain an internal constancy of chemical-reaction rates even in changing external environmental conditions. An example would be our ability to keep a fairly even body temperature during cool spring weather and hot summers. The constancy of the *milieu intérieur* was championed by no less than the great French, nineteenth-century biologist Claude Bernard, a man now referred to by many as the father of physiology and experimental medicine.

At the time I was studying the metabolic rates of developing moths. They are, of course, cold-blooded animals, and thus their body processes are subject to all changes in ambient environmental temperature. But I found that even though I held their body temperatures constant, the rates at which they used oxygen varied broadly. At first I didn't think much about this because biological data vary much more around a mean than the results produced by sciences like chemistry and physics. But then when I plotted what I assumed would be just a scatter of data points I noticed a trend, and when I superimposed a trend line over the points I found a rhythm. Rhythmic homeostasis (if you will) was the condition in this moth rather than Claude Bernard's straight-line version.

I rushed to the library and read everything I could about biological rhythms, which, back then, was an easy task because that literature was sparse. But I became a believer and, as such, took on the same problem that others already in the field were facing: trying to convince colleagues of the reality of organismic rhythms. Locally, I had few supporters. The usual response to our claim was, "If you hold the organism's temperature constant, his physiological processes will be constant—not rhythmic." My answer, as a charter zealot of the Anti-Claude (Bernard) cabal, was, check your data again.

Finally, with our unyielding persistence, nonbelievers eventually became convinced of the reality of rhythms, and the search was on for the clockworks that generated them. Feeling that a unified effort would give a boost to the investigation, in 1960 the first international conference on the subject was held at the famous Cold Spring Harbor Biological Laboratory on Long Island, New York. Experts from all over the world, plus a handful of tyros like myself, assembled and presented their experimental data and the hypotheses they had drawn from them. I was enthralled by the whole proceeding, but was also discouraged: After listening to the experts, in my naiveté I felt sure that the clock would be found before I even finished graduate school. That was the first of the many mistakes I would make during my career as a biologist. Contrary to my belief, successful searches proceeded very slowly, and the Nobel Committee has not even begun to consider a chronobiologist for the prize (but it will).

## CLOCKS FOUND

You readers are now cognizant that a single-cell organism can house a fully sophisticated living clock. However, most of the early investigators worked with multicellular plants and animals and therefore tended to focus on their favorite research subject in their search for the clock. Thus, what follows are the results of investigations using multicellular organisms.

Often, pioneering research on a problem produces very dramatic results. This is because in a first excursion into the unknown, uncomplicated experiments can be performed that often yield clear-cut, easily reproducible, fundamental results. Subsequent work on the project often just produces small details supporting and refining the original finding. Early living-clock investigations searched for the anatomical residence of the timepiece. The following are a few of the most clever and rewarding explorations.

## The Silkmoth Horologue

During the development of the *cecropia* silkmoth from egg to adult, it passes through worm-like larval and pecan-shaped pupal stages. The pupae, nestled in their cocoons, rearrange their molecules into a packaged version of the adult form, and when the task is completed—just as we saw with the fruit fly (Chapter 7)—the pupal case bursts open (eclosion) and (unlike the fruit fly) a beautiful winged-adult emerges. The unvarying appearance at dawn of this particular species suggested that eclosion was under the control of a pupal clock, and studying the process under constant conditions confirmed this supposition. Surgery was chosen as the means to search for the clock. The pupa was opened up and various parts were removed to see what would happen. Pioneering work must rely on trial and error, and success came slowly. Finally, in desperation, the entire brain was removed. As outrageous as this may seem, the undaunted, mindless pupae went right ahead and eclosed, but now the adults emerged at all times of the day. When the brain was replaced, even if it was inserted into the wrong end, for example, into the pointy abdomen, the emergence rhythm with a dawn phase was reinstated. Clearly, the brain was serving as a gate controlling eclosion, but was it actually the clock, or did it just serve as a middleman between the clock and the eclosion process? One thing was certain: Because the brain carried out its role without being directly attached to the rest of the nervous system, it had to be exerting its influence hormonally, rather than via nerve pathways.

The next question to be answered was what part of a moth's body sees the light of day and thus sets the clock to the proper eclosion time. A brainless pupa (the brain was reserved for later use) was inserted snugly into a hole in a box, with its head inside and abdomen outside the box. Outside the box (right side of Figure 9.1) was the normal day/night cycle; inside the box (left side of figure) was an artificial, reversed night/day cycle. Thus, when it was daytime on one end of the pupa it was night on the other end. When a brain was inserted into the end exposed to the normal day/night cycle, emergence occurred at normal dawn. If a brain was introduced to the head end inside the box, the adult emerged at box-time dawn, which was normal nighttime outside the container. This means that the brain functions as the eyes, it is the phase-setter for the hands of the clock.

Jim Truman, then at Harvard, the human brain power behind these experiments, carried his moth-mutilation manipulations a step deeper. In this endeavor he had help from another species of silk moth, one called *pernyi*. This moth also underwent an eclosion rhythm but it differed in that emergence ensued at dusk, rather than at dawn. Brain exchange between species was the order of the day: *Pernyi* brains were transplanted

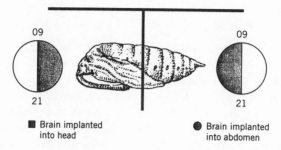

*Figure 9.1. A brainless silkmoth pupa with its head and tail ends exposed to different light/dark cycles.*

into brainless *cecropia* and vice versa. The recipients were placed in a normal day/night cycle. The pernyi-body/cecropia-brain chimera eclosed at dawn, while the cecropia-body/pernyi-brain combo emerged at dusk. This neat trick made it clear that the brain clearly contained the clock and was not simply a relay station between a real clock and the eclosion process.

Both of these moths display flight rhythms as adults. This rhythm is also under the control of the brain, so if the brain is removed the rhythm disappears. However, if after removal the brain is inserted into the abdomen the rhythm does not reappear, meaning that brain hormones are not fundamentally involved with this rhythm. The brain, in this case, must be attached to the rest of the nervous system for the rhythm to reappear, meaning it broadcasts its timing information via nerve pathways.

While on the subject of insect brains as clocks, I'd like to digress for just one paragraph back to the fruit fly brain. Its removal destroys the fly's activity rhythm. Two lionhearted chronobiologists decided they would transplant a brain to the abdomen, just as had been done in silkmoths. Stop reading for a moment and reflect on what a hassle that endeavor would entail. The brain is no larger than a fly speck; getting it into a tiny abdomen would be only slightly less arduous than getting it out of the even smaller head without decapitating the animal. The fly not only had to survive such drastic surgery, it had to be able to locomote normally afterwards. Out of 55 tries the surgeons succeeded only 4 times. But it worked, in these four the lost rhythms were restored. Thus, unlike moth activity, the fly brain exerts its influence on activity via hormones.

*A Search for the Elusive Crab Clock?*

The clocks in *cecropia* and *pernyi* moths exerted their temporal gating mechanisms on eclosion via hormones. There is a suggestion in the literature that the same may be true in some crabs (moths and crabs are related arthropods). The following account describes a way of testing this idea.

Crabs have no necks and cannot turn their head ends around to see what is behind them. This limitation is somewhat rectified by having their eyes mounted at the tips of movable stalks that can be pointed in many directions (crabs run sideways and those that are blessed with long stalks have the potential to point one eyestalk to starboard and the other to port so they can simultaneously see where they are going and where they have been.)[10] The stalks do double duty in that built into them are hormone producing glands. These secretions unquestionably function in crabs' day-to-night color-change rhythm, and in one species there is also a suggestion that the eyestalk hormones may play a role in staging the animal's tidal-activity rhythm.

When I first learned of this suggestion I examined the possibility using the crab I happened to be working on at the time, the penultimate-hour crab, an animal with a very interesting activity rhythm. To stop eyestalk-hormone production was simple, I simply cut off the entire stalks at their bases. While the amputation did produce changes in the animals' behavior, I found no indication that stalk secretions were involved with the animals' activity rhythm. However, other parts of a crab's innards also produce hormones, and I fantasized for a brief moment about attempting the required delicate and difficult operations of excising, or electro-surgically burning out, these areas. But the time required and my fat-finger lack of operating skills discouraged such an approach. An easier way came to mind.

But before presenting this revelation, I must pass along two required background facts about crabs. As mentioned briefly before, crabs have ten spider-like legs sticking out from their bodies; predators, at the onset of their attack, often grab a leg as the most convenient handle. This is so common that natural selection has provided at the base of each crab leg a pre-ordained breaking plane. Under attack a crab can voluntarily break off a grasped leg at that point, and while the predator stops its assault to consume the leg as a preprandial *hors d'oeuvre* before the main course, the nine-legger runs off to safety. With subsequent molts the missing leg is replaced. If one should ever want a legless crab, pinching each leg with a pair of pliers would cause them all to be cast off.[11]

The other piece of information you must have is this. A crab's circulatory system consists of only a heart and two short blood vessels. The blood is simply pumped out into the body spaces where it bathes the organs for a while and eventually sloshes back to the heart to be pumped out for another round trip. During this helter-skelter journey the blood delivers food, oxygen, and hormones to the cells and picks up wastes for excretion. It is a very simple system, but quite adequate for an organism without a particularly physically active lifestyle.

Finally, the background information now complete, on to my experiment. I kept one set of crabs in constant conditions in the laboratory until I was sure that their activity rhythms had stopped. I then collected a set of fresh crabs and tested them to make sure they had nice rhythms. Next, I carefully cut a hole, slightly smaller than a dime, through the shelled backs of both groups. I held a rhythmic crab and an arrhythmic one back-to-back, hole-to-hole, and glued them together with sealing wax. Now the blood of both crabs began to freely mix, and none leaked out. Next I applied pliers pinches to each of the legs of the rhythmic crab and all were cast off. The finished product is seen in Figure 9.2. Although the procedure appears demonic, apparently for the crabs it was not: all survived for more than a month in the lab; the bottom one ate for two, the eyestalk of the piggyback crab actively scanned its new upside-down world, and the bottom one walked around without noticeable difficulty. Each of these parabiosised (as such a union is called) pairs were placed in activity recorders (like those described in Chapter 6) and observed in constant conditions. Only the arrhythmic crab had legs, meaning that any activity documented had to be accorded to it. The pattern of activity should show no rhythm, unless a rhythm-causing hormone secreted from someplace in the innards of the legless rhythmic crab riding on top entered the walker's blood. No rhythm was produced in any of the parabiosised pairs.

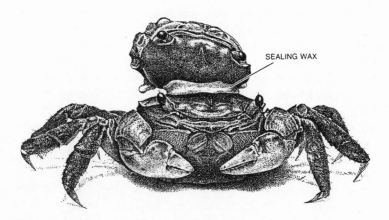

SEALING WAX

*Figure 9.2. Two penultimate-hour crabs parabiosed back-to-back. The one on top is legless.*

Thus, at least in the penultimate-hour crab we have evidence that hormones do not drive activity rhythms.

When the experimentation was terminated, all of the conjoined pairs were alive and active when I released them back onto their home shoreline. I have no idea of what goes on in a crab mind, but as I freed them I could not keep myself from trying to imagine how I would respond if I were a free-living penultimate-hour crab getting my first look at one of these strange double deckers that just landed in my little part of the universe. They would be too large and ominous for me to attack, and too large to fit in my burrow, meaning mating was out of the question. Probably I would just go up to one and say, "Whatever the aliens were looking for, I hope they didn't muck up your clock." As a human making the same startling sighting, I'd go home and get out of my wet clothes and into a dry martini.

## The Cockroach Brain Clock

Those of you who have ever lived in an apartment complex in a large city must be fully aware that your nearest neighbor is virtually always a roach. A sound sleeper may not be aware of this, but those who enter the kitchen at night for a glass of water or a snack instantly learn that roaches are nocturnally active: When you snap on the light your kitchen floor could well be a cockroach carpet (a single female can produce 35,000 descendents in a year). If, instead of stamping on as many of them as possible before they escape back into protective crevices, you were to collect a few for study in constant conditions, you would find that they have fine persistent activity rhythms.

This unloved animal, this immortal enemy of frustrated exterminators, also has served as an important research subject in the quest for the anatomical location of the living clock. Neurosurgery has been the approach of choice, in spite of the incredible skill required to open a roach's tiny armored head; the dexterity needed to make knife cuts across various areas of the even smaller brain; and finally the difficulty of closing the incision so the patient with an open circulatory system like a crab's does not bleed to death. For the procedure to be adequate, the roach must live for weeks to months after the operation and be able to walk normally so its activity rhythm can be studied. Using this somewhat random, Zorro slashing approach, it was finally found that if a cut is made between structures called the optic lobes and the rest of brain, the locomotor rhythm is destroyed. There are two optic lobes, one on either side of the brain, and if only one of the lobes is separated from the brain, the rhythm persists, demonstrating that there are two, redundant, master clocks in this animal.

If the severed lobes are left in place in the animal, after three to five weeks the lost rhythm reappears; microscopic examination reveals that the nerve fibers between the two lobes and the rest of the brain have regenerated. This finding set the stage for the next experiment.

When studied in constant darkness, the locomotor rhythm of one cockroach strain has a period length of 22.7 hours, while the period of another strain is 24.2 hours. The experiment was an obvious, but difficult, one: The optic lobes of long-period animals were removed and replaced with short-period lobes, and vice versa. After a four-week wait required for the lobes to establish neural connections with the brain, the recipients' rhythms returned and expressed the new periods dictated by the inserted foreign lobes. This was the cinching result: Clearly, cockroach timing resides in the optic lobes.

Lastly, the *pièce de résistance:* One short-period lobe was used to replace one of the lobes in a long-period animal. During the weeks while neural connections were being established between the introduced lobe and the brain, the animal's activity rhythm was still governed by the original resident lobe so the period of the activity rhythm was long. Somewhat surprisingly, the long period continued even after the new nerve connections with the short-period lobe were completed; somehow, the long-period lobe was able to suppress the action of the transplanted short-period lobe. That being the case, next the junction between the long-period lobe and the brain was severed, and this permitted the short-period lobe to express itself and produce a short-period rhythm. Again the investigator waited, letting the nerve fibers grow back between the long-period lobe and the brain. When this was complete, a long period appeared in the activity rhythm, but the short period remained also. Now the animal's persistent locomotor rhythm boasted both long and short periods simultaneously! This special result is shown in Figure 9.3. Each dash represents an hour of no activity. Each black bar represents eight-hour-activity bouts (if the actual rhythm had been plotted, the dark bar would be the activity peak). If this animal had just possessed a simple one-peak-per–24-hour-day rhythm, each daily black bar would have fallen exactly beneath the previous day's, producing in the drawing something similar in appearance to a single stack of poker chips. But in this animal there are two rhythms running simultaneously, and thus two activity bouts are produced each day: the long-period rhythm, whose peaks come later each day (when plotted on a 24-hour scale, as in Figure 9.3), is represented by the diagonal line of bars scanning to the right; the short-period rhythm, where activity bouts occur earlier each day, is represented by the diagonal bars moving to the left. Where the two rhythms cross (the dark diamond in the center) the roach activity is maximal because both

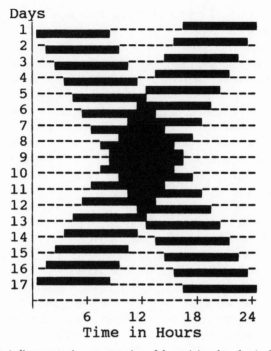

*Figure 9.3. A diagrammatic representation of the activity plot of a singly cockroach displaying simultaneous short- and long-period rhythmic components. See text for details.*

clocks are mandating full power. Terry Page a Prof. at Vanderbilt University and the orchestrator of this experiment, clearly managed to create two-part harmony in one roach.

*The Mammalian Suprachiasmic Nuclei*

In 1967, Curt Richter, a fine psychobiologist and a member of the National Academy of Science, had a laboratory full of wild rats gleaned from the alleys and sewers of New York City. Rats, as you who are unfortunate enough to know, are masterful at avoiding poisoned bait. Thus, in the service of his country during World War II, Richter's job was to elaborate an effective, foolproof way to exterminate rats in case the Axis developed rat-borne germ warfare. His mission accomplished (meaning he learned how to eliminate them, but, as New Yorkers know so well, he was never hired to serve as a Gotham's Pied Piper), he became fascinated with rats' rhythms and for several years tried to learn where in their bodies the living clock resided. The brain, which reigns over most of the show in higher animals, was his number-one suspect.

His approach capitalized on the bold stroke: He opened their skulls, made a single knife cut in various areas of the brain, or removed parts thereof, closed the wound, revived them with mouth-to-snout resuscitation (as is rat physiologists' wont), and observed whether their rhythms persisted or were lost. After more than 200 operations he found that lesioning the hypothalamus destroyed the rat's rhythmicity.

The hypothalamus is a small mass of brain tissue below the cerebrum and just above the pituitary. It has many functions including the synthesis of some hormones and regulating the release of certain hormones from the pituitary.

Subsequent clever work in two other laboratories eventually traced what appeared to be a clock in the anterior end of the hypothalamus just above the optic chiasma (the place where the optic nerves cross so that tracts from the right eye go to the left side of the brain and vice versa). Here in the rat are two clusters of about 8,000 cells each, which together are called the supra (meaning above) chiasma (the intersection of optic nerves) nuclei (in this case meaning a group of nerves). Put them all together they are verbalized as suprachiasmatic nuclei. Happily, all but the punctilious and pedantic refer to them simply as the SCN. If the two of them are destroyed, most, but not all, of a rat's rhythms cease.

But destruction followed by loss of rhythmicity does not guarantee that the SCN are clocks, or, the clock. The argument goes something like this. If one turns the ignition key in his car and nothing happens, it does not necessarily mean that the battery is dead. It could be that an ignition wire between key and starter is damaged. If an area of the brain that is suspected of being the clock is destroyed and a rhythm elsewhere stops, one cannot be sure the demolished spot was not just a relay station between the actual clock and an organ elsewhere it caused to be rhythmic.

Conflicting evidence soon turned up. Even after the destruction of the rat SCN the rhythm in oxygen consumption of the liver persisted, as did the rhythm in hormone secretion by the adrenals. In fact, these rhythms even persist after the organs have been removed from a rat and maintained alive in culture. The liver and adrenals obviously have their own personal clocks.

Several subsequent experiments elucidated a fuller picture of what was going on here. If the SCN of a rat was ablated, but a new SCN from another rat transplanted in, with time the animal's missing rhythms returned. There is a mutant hamster whose activity rhythm is only 20 hours in length in constant conditions. When its SCN was replaced by a SCN from a hamster with a 24-hour rhythm, the longer period was the one then displayed by the recipient. Furthermore, if the recipient's SCN was not removed, the result of the transfer was the presence of both periods being displayed simultaneously in the host's activity rhythm: One peak formed every 20 hours and another at 24-hour intervals.

The SCN are an intimate part of the brain and could be spewing out timing information via impulses along their nerve fibers, or by secreting chemical signals, or possibly by both means. In a struggle to answer this question, SCNs were removed from hamsters and enclosed in a special capsule that permitted the molecules they synthesized to diffuse outwards (and nutrients to move inward to keep them alive), but prohibited nerve fibers from developing between the SCN and the hypothalamus. When a packaged SCN was inserted into an animal whose SCN had been ablated, the animal's lost rhythms reappeared. This trick demonstrated that it is the secretions liberated from the SCN that control the animals' rhythms, without the help of nerve-fiber transfer.

A second way of demonstrating this was to simply inject a slurry of loose SCN cells into one of the brain cavities of an animal whose SCN had been ablated. The previously sans-SCN animal became rhythmic again.

There are even more ways to demonstrate that the SCN are really autonomous clocks, such that they are able to measure off time on their own. When electrodes are implanted into them, their electrical-activity output is recorded as a series of spikes appearing on an oscilloscope screen. The frequency of the spikes varies over the interval of a day in a rhythmic fashion: They are more rapid during the day than at night. Next, SCN are removed from a brain and kept alive in a special culture medium; the spike-output rhythm still persists, confirming that the isolate is an independent clock. One step further along in this experimental protocol, SCN cells are teased apart and maintained in a life-support system. When electrodes were attached to individual cells it was found that their rhythms persisted, meaning that each cell in a SCN is a sovereign living clock.

So where does this leave us; remember we still have to deal with the oxygen-consumption rhythm that persists in liver tissue isolated in culture and the hormone-production rhythm in isolated adrenal glands. Both these organs obviously have their own clocks. The latest thinking is that the SCN function as the primary clock—the master horologe if you wish—that causes some physiological and endocrinological processes, plus behavior such as locomotion, to be rhythmic. Their other function is to oversee the phase of slave clocks like those in the liver and adrenals, keeping all the rhythms of an individual in tune . . . so to speak. Either removal of the SCN from the rat, or removal of the liver and adrenals and keeping them alive in culture, weans those organs' clocks from the slave master.

One last bit of information must be communicated. The SCN clocks are set by the ambient day/night cycle, and the eyes are the receptors of this lighting information. The messages received are carried via optic nerves directly to the SCN (not via the nerve tracts involved with vision, but by separate tracts dedicated to that specific task). In fact, the SCN were originally discovered by the very difficult, time-consuming task of following, histologically, these tracts inward from the eye to their terminus in the brain, which turned out to be the SCN. The discovery was rather similar to the rewarding endeavor of following a map trail to the spot marked X, and there unearthing pirates' buried treasure.

All mammals have SCNs, including humans. Ours consist of about 45,000 cells and are twice the size of a rat SCN, but are still much smaller than a BB shot. They are similar in organization to those of other mammals, but contain two large neuronal populations that are not as common in other species. It is not known if our SCNs serve us as a clock because it would be unethical to ablate human SCNs to see if some of our rhythms ceased. But philosophically there is little reason to presume that they are not clocks if we accept as fact that we too are a product of evolution. By the eighteenth week of gestation SCNs have formed in the human fetus. Several rhythms have been described in fetuses, but it is unknown whether they are driven by a fetal clock, or if they are just a result of the dictates of a maternal clock passed via the umbilical cord.

The rhythms of some blind people can still be set by the day/night cycle. The explanation usually given for this is that these people are blind because visual information cannot pass over the neural tracts that lead to the vision centers in the brain, but moves instead via the special nerves leading to the SCN (which are still functional) so the clock receives the status of the changing-light regimen outside the body. Or, maybe light falling on the back of the knee (introduced in Chapter 4) will set these clocks; we should maintain a wait-and-see attitude.

*A News Flash about the SCN*

When the foot soldiers of the Fourth Estate became aware of the existence and function of the SCN, almost overnight biological rhythms became big news. Living in a vociferously anthropocentric me generation, discoveries in lowly rats were not worthy of media hype, but reporters made the inductive jump and banner headlines blared, "HUMAN BODY CLOCK FOUND." And, in this new age of misinformation, the thrust of most stories offered by the popular media since then state that the search is over: The suprachiasmatic nuclei are the tick tocks of life. But of course, the reporters do not know what you readers all know: Rhythms and clocks exist in single-cell organisms, in the entire plant kingdom, and in all the invertebrate (animals without backbones) species . . . and not even one of them has a SCN!

## The Survival Value of a SCN Clock

The possession of any sort of living clock fundamentally determines just when an animal will be active (for example, during the day or during the night; during the high or during the low tide; at dawn or dusk; et cetera. Timing can be very important: For instance—if a fiddler crab were to emerge from its burrow at high, rather than low tide it could become a crunchy tidbit for a stripped bass. This survival interpretation of the clock's value seems obvious, but until fairly recently, it had not actually been tested.

In the Mojave Desert in southeastern California lives the white-tailed antelope ground squirrel, a little rodent about eight inches long with a white stripe on each side of its body. Other than a few saltbushes and creosote bushes scattered around the desert floor where the animal lives there is little to obscure an observer's view of the animals' daily wanderings around its home territory. Thus the squirrel is known to have a good diurnal active rhythm. At nighttime the squirrel resides in the relative safety of its burrow. Rattlesnakes and owls are common, potential, nighttime predators.

A small number of these squirrels were first studied in running wheels in the lab to ensure they had robust activity rhythms, and then their SCNs were surgically destroyed. This, of course, caused the animal to become arrhythmic in constant conditions, but otherwise was not disruptive to a survivor's health so long as the animal is kept in the laboratory. But what would happen if it was returned to a natural setting was unknown. So, in the first experiment to find out, five squirrels with their SCNs decommissioned, along with seven unoperated animals, were placed in an outdoor compound with a low fence surrounding it to keep the local rattlesnakes from slithering in for a meal. Artificial underground dens were provided and were quickly occupied by all the squirrels. But after a few nights, as caught on video, it was not a snake who visited; it was a feral cat that jumped the fence, and waited with the immobile patience so characteristic of its much bigger brethren. Each time a squirrel would foolishly pop its head up out of its hole, the cat would pounce. I'll forgo the grizzly part and just say that by morning when the feline was shooed away, it had taken three of the SCN-lesioned animals and two of the unoperated ones.

While the numbers here are too small to draw any conclusion, or certainly to reach any statistical validation, there is at least the impression given by the body count and video footage that the clockless animals stuck their heads up more frequently than the others that night and got picked off. There is thus a suggestion that the clock provides the proffered survival value.

This experiment was just a pilot study and the results far less exciting than a cavalry charge. A larger study is now underway, but I must stress how difficult this kind of research is to do. To begin with, one must be a capable brain surgeon on small subjects, meaning that the sans-SCN patient must not only survive the operation, but must remain as normal as possible. The next problem is that the experiment is conducted out-of-doors meaning any violent weather and earthquakes can destroy it; that an observer must always step over, rather than on, rattlesnakes; that an area can be found that is free of all-terrain-vehicles (high gasoline prices may help here); and that one can devote years to the study if necessary. Fortunately, there are a few chronobiologists willing to risk all this; the others of us not only look forward to learning what they find, we invite them over occasionally for a good meal and a chance to sit indoors for an evening of sociable persiflage.

## The Pineal Clock

The pineal has an interesting history, spanning the disciplines of philosophy, biology, and medicine. As mentioned previously, after its initial discovery in humans no function could be found for it, so Descartes decided it must be the seat of our soul. We have since learned the pineal's several functions, but still have not pinpointed the soul's residence.[12]

A pineal is present in most vertebrates, but its anatomical position and function differ from species to species. In fish, amphibians, and reptiles it sits on top of the brain just under the skull. In frogs and lizards it responds to light and is known as the third eye. In birds and mammals it lies deeper within the brain and functions in making and secreting the hormone melatonin. In humans, it is a pea-sized organ shaped like a pine cone (hence its name) perched on top of the so-called brainstem (a position inspired by evolution) near to the hypothalamus. It is no longer close to the skull as it has become covered by our large cerebral hemispheres. It was the last endocrine organ to be discovered in humans.

Early work on the sparrow cleverly demonstrated that the pineal functioned as a living clock. If a fully sighted sparrow is put into constant conditions both its activity and body temperature undergo a daily rhythm. After the skull is opened and the pineal destroyed, or removed, the rhythms are lost. But if a pineal from another bird is then implanted in the eye (easier than opening the skull again) the rhythm quickly returns. Clearly the pineal clock exerts its effect hormonally, rather than making new neural attachments through the eye into the brain.

When a potential pineal-donor sparrow is placed in a reversed day/night cycle (the same lighting schedule as if the bird had been moved to Chungking, China) for several days, its rhythms are reversed. Its pineal is then transplanted into a bird that had been kept in local day/night conditions before its pineal had been removed. The recipient (now maintained in constant conditions) becomes rhythmic again, but the phase has been reversed, in that it is now what the donor's phase had been.

As mentioned above the pineal exerts its timing information via the hormone melatonin, which is secreted into the blood for transport around the body. Melatonin is made by the conversion of a compound called serotonin in the following reaction:

$$\text{Serotonin} \xrightarrow{\overset{\text{N-Acetyl-transferase}}{\text{NAT}}} \text{Melatonin}$$

NAT is an acronym for a particular enzyme (=catalyst) with a very long name that speeds up the reaction to a point where easily measurable amounts of melatonin are produced. A simple biochemical analysis for NAT activity done at regular intervals throughout the day has demonstrated that it displays a fine daily rhythm. If a way could be developed to keep an isolated pineal alive in culture, and then follow the changing activity of NAT, the autonomy of the pineal clock could be insured.

A way was found to do this, but not with the sparrow pineal, which, as would be expected by the small size of the bird, is really minute. Instead, the much larger chicken pineal was used. Removed chicken pineals were kept alive in culture and found to produce NAT in a rhythmic fashion for a couple of days in constant conditions. While isolated in culture the phase of their rhythm could be altered by exposing them to different day/night cycles! The chicken pineal is thus a bona fide autonomous clock spewing out melatonin in the form of a well described daily rhythm.

After experimenters increased their culturing and micromanipulation skills, they began shaving slices off isolated chick pineals and found that one whittled down to only one-eighth of its normal size still continued to function as a clock. Eventually, these investigators were able to separate pineal cells one from another, and maintain them in culture; still, they continued to be rhythmic. The same story again, each cell is provided with a living clock.

Returning to the sparrow, it was clear that the rhythmic output of melatonin from its pineal was the controlling factor of an individual's activity rhythm. If melatonin was injected into the bird, it disrupted the activity rhythm. That information coupled with the fact that pinealectomy caused a total loss of rhythmicity cinched the fact that the sparrow pineal was a clock.

But then came a surprise, the results from what is sometimes called one experiment too many. Removing the pineal from chickens did not cause their activity rhythm to destruct! To make a long story short, it was eventually found that the chicken's SCN controls its rhythms.

Since then it has been found that some birds have a clock in their eye. The importance of these three clock types varies among bird species.

### A Molecular Clock—The Dernier Cri

Background information: Within the cells of all higher organisms is a nucleus, and within it are chromosomes. Each chromosome is made of protein and a molecule of DNA (deoxyribonucleic acid). As molecules go, DNA is a very large one, so large in fact that it can be seen using an electron microscope. In spite of its large size, and its extreme importance to living things, DNA is a relatively simple molecule: It is mainly made of just four, small, different molecules called nucleotides that repeat their presence in various combinations over and over again in a linear fashion. The DNA molecule is shaped like a typical ladder that is twisted into a helix (or, for pasta aficionados, the shape of rotini). Its structure was worked out by James Watson and Francis Crick, who, along with Maurice Wilkins, shared the Nobel prize for their brilliance. After receiving this honor, Crick said, "I was almost totally unknown at the time, and Watson was regarded, in most circles, as too bright to be really sound!"*

Each DNA molecule is a strip of consecutive genes. Each gene consists of varying number of nucleotides and is distinguished from other genes by the order in which the four types of nucleotides are bound together. Genes vary in length from just a few nucleotides to thousands. To make this personal, each human cell contains a passel of roughly 3.1 billion nucleotide pairs. I have read statements claiming that if they were arranged end to end and presented as lines of a text they would fill 800 Bible-length books (a silly comparison, but it does drive home the relative magnitude: the genome string is much larger than most individuals' savings accounts). Human DNA consists of something like 25,000 to 35,000 genes, a number that surprised most involved in deciphering the genome: Up to then the generally accepted estimate had been 100,000 genes.

Most all the basic information needed to run our lives exists in the nuclear genes, but the majority of the activity needed to keep us alive and functioning each day takes place in the cell cytoplasm outside the nucleus. Using a factory analogy, decisions are made in the main office and are then carried out to the shop floor where the workers make widgets and doohickies. The same arrangement obtains in the cell: Orders from the DNA are transported by special molecules, called messenger ribonucleic acid, or mRNA for short, out to the cytoplasm where they are used to instruct the construction of a whole host of different, specific proteins. The proteins form many of the structural elements of the cell and also serve as enzymes in the cell chemistry.

Whenever the same gene is activated, the same proteins are subsequently built outside the nucleus, unless something has happened to the molecular structure of the gene. Genes can be intentionally altered in the laboratory by exposing cells to X-rays, UV radiation, or certain chemicals called mutagens. When altered in these ways—or as a result of a natural accident—a gene is then referred to as a mutant.

To summarize in one sentence: Genes in the nucleus make mRNA that diffuses out to the cell body and causes the synthesis of specific proteins that are the basis of most aspects of cellular metabolism. Now back to biological rhythms and clocks.

Begin by recalling the fruit fly eclosion rhythm described in Chapter 7. Two thousand flies were subjected to a mutagen called ethylmethane sulfonate, and the rhythms of each of their offspring were examined for abnormalities. Of this enormous group, only three mutants associated with rhythms were found. In constant conditions the period of the eclosion rhythm of one fly had lengthened to 29 hours, another had a period of 19 hours, while the third had become arrhythmic. Remember that the normal period of the eclosion rhythm in constant conditions is ~24 hours. Further testing showed that a single gene had been altered and it was responsible for the different changes seen in all three flies. The gene responsible was named period, abbreviated as *per*. The long-period mutant was called $per^1$, the short mutant, $per^s$, and the arrhythmic one $per^0$. The normal gene—the unaltered version—is referred to in shorthand as $per^+$.

Genes, you will remember, first pass their directive to mRNA, which moves out of the nucleus and organizes the construction of specific proteins. In this case that protein is called PER (signified by using all caps) so it is easy to remember it is a product of the *per* gene. Subsequent studies demonstrated that both the *per*-mRNA and PER, oscillate in abundance in a 24-hour pattern, the former leading the latter as would be expected.

Genetics has become very sophisticated over the last couple of decades, and several years ago it became possible to remove a *per* gene from one fly and put it into another. The trick first involves borrowing a relatively simple strand of nonchromosomal DNA, called a plasmid, from a bacterium. The plasmid is removed from the bacterium, a $per^1$ gene is inserted into it, and the plasmid is then injected into a $per^0$ fly embryo. The embryo is allowed to grow into an adult that mates, and whose offspring are tested for rhythmicity. If the transfer of the $per^1$ gene is unsuccessful, the genetically $per^0$ adult would, of course, remain arrhythmic. If the transfer is successful, the adult will have a long-period rhythm. As an aid to quickly recognizing a viable transfer of the $per$ gene, the investigators added another fly gene to the bacterium plasmid, one that causes fly eyes to be a rosy color. Thus, when the above experimentals reach adulthood, all the investigators had to do is pick out the rosy-eyed flies to test for rhythmicity. The offspring had long periods. All-in-all, a very clever manipulation.

To locate where in the fruit fly the period gene resides, a bacterial plasmid was again exploited. This time $per^+$ and a gene, which made a product that would turn blue in the presence of a specific test chemical, were combined in a plasmid and inserted into a $per^0$ embryo fly as above. When the fly eclosed into an adult and mated, its now-rhythmic offspring were cut into thin sections and flooded with the test chemical. Anywhere the blue color appeared signaled the presence of a $per^+$ gene. The result: Numerous tissues in the fly turned blue, indicating a wide distribution of clock-potential.

This ubiquity of $per$ throughout the body cells was then demonstrated in another way. This technique was a step upwards in that it allowed one to see whether $per$ was producing its product or not in a living fly. The luciferase gene (described in Chapter 8) was inserted onto the fly in such a way that when $per^+$ was turned on, so was luciferase production. The flies were fed luciferin, and each time the $per^+$ gene became active the flies also glowed like a firefly. The $per^+$ gene turned on and off on a 24-hour basis. Next, the flies were cut into three parts: They were decapitated and detailed (had their *derrières* lopped). The rhythmic expression of the $per^+$ genes in head, thorax (the middle piece), and abdomen all continued to glow in unison for several days. This finding and the previous one certainly are consistent with what has been found before in other organisms: Almost every body cell contains its own clock(s). Together, the above experimental results make it clear that the *period* gene is at the very least an intimate part of the living clock.

Recently, several other fruit fly genes have been identified and demonstrated to work together with $per$ to make what appears to be a living, molecular, timing device. Instead of outlining the experimentation—which is a rather recondite tale—leading to the discovery of each gene, I'll just name the basic genes and then describe how they function together. Even this brief and superficial account will probably be more information than many readers will want, but don't miss the take-home message: The world now has a good idea of what is going on at the molecular level of biological timing. In the fruit fly the four main genes are: period (*per*), clock (*clk*), cycle (*cyc*), and timeless (*tim*). Together (with the involvement of some players that I will leave unnamed) they form a feedback loop (a series of reactions that can be expressed as a ring). Because the equation is circular—like the chicken-or-egg conundrum—I can arbitrarily start the description at any point in Figure 9.4. Follow the arrows. Two genes in the nucleus, *clk* and *cyc* (in boxes), construct their special messenger RNAs (not shown) that move out of the nucleus into the cytoplasm where each directs the synthesis of its specific protein, CLK or CYC. Next the two proteins eventually combine making CLK-CYC molecules. This product then moves into the nucleus and combines with a special segment of a chromosome called a promoter, which then activates the adjoining *per* and *tim* genes (also in boxes) to synthesize their messenger RNAs (also not shown). These messengers move out into the cytoplasm and direct the synthesis of their particular proteins, PER and TIM, that slowly accumulate. With time

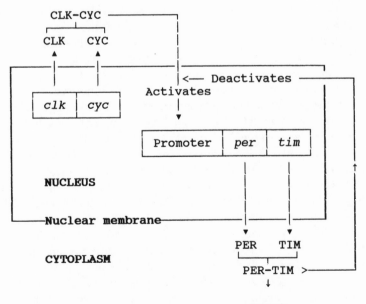

*Figure 9.4. A schematic of interactions of some of the genes and their products that form the fruit fly living clock.*

these two molecules combine into PER-TIM, which, when sufficiently abundant, eventually feeds back into the nucleus and inhibits further stimulus of *per* and *tim* by CLK-CYC. As a consequence of this, PER-TIM shuts down its own production, and those PER-TIM already present spontaneously degrade. With PER-TIM no longer available, *clk* and *cyc* again become active. Then the whole molecular-peregrination cycle—which has taken about 24 hours to complete—starts over again. This then, according to the molecular geneticists who deciphered the loop, is the clock (scientists, while not with certain politicians, there is no need to deconstruct the meaning of is to get to the fact of the matter). Among several more things that we now need to know, is how this circle of events hooks to the myriad of other cellular process making them rhythmic. For the time being we must wait for the answers, giving me a chance to say time will tell.

The genomes of many other organisms have been totally or partially worked out. It is known that the fruit fly has only half the number of genes as humans. Mice and humans are separated by only about 300 genes. The genomes of men and chimpanzees differ by only 1.7 percent! Given data like this readers should now be questioning: Do humans, animals other than the fruit fly, and plants have *per* genes? The answer is yes, in a way. *Per* genes in other organisms share many of the same nucleotide sequences, but not necessarily all of them. Below is a comparison of the *per*-genes products—which reflect the gene order that produced them— of humans, mice and fruit flys. The PER products of humans (hPER), mice (mPER), and *Drosophila* (dPER) are 1,290, 1,291 and 1,218 amino-acids long, respectively. To give the reader a feel (in a limited space, the segments shown are just 5 percent of the complete molecules) for how similar they are to one another I have copied here just 60 amino-acid segments (actually only 59 for *Drosophila*). These compare just one homologous region between the three molecules (the different amino acids are represented [and coded] by a single capitol letter). The asterisks along the top indicate identical amino-acid locations in humans and mice. The arrowheads (▲) below point to *Drosophila* amino acids that also occur at the same locations as in hPER and mPER. In this example, the amino acid sequence in humans and mice is identical (the most common agreement), but the fruit fly sequence matches them only 40 percent of the time. The large difference in the latter would be expected because humans and mice are closely related, while the fruit is a distant evolutionary relative.

Some other organisms in which *per* genes have been identified are chickens, spinach, rape (a mustard-family plant we use as a source of canola oil), and Venus' flower basket (Chapter 5), to mention just a few. Interestingly, in Venus' flower basket the *per* genes are apparently not in the nucleus, but instead are in the chloroplasts. Chloroplasts are little or-

```
********************************************************************
hPER  LAGQPFDHSPIRFCARNGEYVTMDTSWAGFVHPWSRKVAFVLGRHKVRTAPLNEDVFTPP
mPER  LAGQPFDHSPIRFCARNGEYVTMDTSWAGFVHPWSRKVAFVLGRHKVRTAPLNEDVFTPP
dPER  TAGASFCSKPYRFLIQNGCYVLLETEWTSFVNPWSRKLEFVVGHHRVFQGPKQCNVFE-A
      ▲▲  ▲   ▲ ▲▲   ▲▲ ▲▲   ▲ ▲  ▲▲  ▲▲▲▲  ▲▲ ▲ ▲ ▲    ▲    ▲▲
```

gans scattered throughout plant-cell cytoplasm that carry out photosynthesis. Each chloroplast has a small amount of DNA in it that is used, among other things, to reproduce itself within its host cell. Remember that a Venus' flower basket filament from which the nucleus has been removed can be cut into ten-millimeter-long pieces each of which remain rhythmic. Apparently this is because each segment has many clock-containing chloroplasts within them!

Our journey is over: We have traced the clock through organ systems, to individual cells, and finally down to the DNA of genes. This end point should probably not come as a surprise. DNA is the fundamental orchestrator of the cellular show: Its two most important functions are to oversee cellular metabolism (the conversion of raw materials into cell components) and to enable cell reproduction. To this we now add that these ubiquitous molecules also govern a major portion of the temporal lives of most all living things.

The search for this living clock has been long and tedious because some of the upward steps had to wait for a general understanding in a speciality field to develop to a level where it could be applied to rhythm work. It all began with the careful observational efforts of botanical and zoological field biologists. They had great difficulty convincing other biologists that clock-controlled rhythms even existed. Undaunted, field biologists collected ever increasing numbers of examples of rhythmicity in plants and animals. When their claims finally began to be generally accepted, more biologists, biophysicists, mathematicians, and others[13] became interested and joined in the fray. Ethologists interpreted and quantified behavioral rhythms. Rhythmic organisms were moved into the laboratory and studied experimentally by physiologists in controlled environmental conditions, and it was there that the anatomical locations of the living clock were found in the eyes, brains, pulvini, and in fact, in every cell of an organism. Then came the biochemical struggles to understand the basic chemistry of the clock. And finally the molecular geneticists got into the game and pinpointed the basic clockworks to organized strings of nucleotides. While there is still much to be learned about the clock, such as, what enables it to run at the same speed at different temperatures, how it is coupled to processes it causes to be rhythmic, et cetera, a shared Nobel Prize should be just a few years away. This award should be applauded by everyone when it comes; the public owes a great debt to the hundreds of contributing scientists who worked for many years to produce this understanding of the living clock: Organismic rhythmicity is a basic property of living things, and almost nothing was know about it until about 50 years ago when vigorous research efforts began.

# NOTES

## CHAPTER 1

1. I learned later that students in the University of Bristol's microbiology class were routinely brought down here to collect *Salmonella typhi*—the causative agent of typhoid fever. Today, in hindsight, I ponder the additional fate of a face full of that mud if one's bungee cord parted, but I suppose the broken neck would preclude that worry.

## CHAPTER 2

2. Yes, beginning by the tenth week of development, the fetus has begun breathing movements. However, it is, of course, not air but amniotic fluid that is breathed in and out. Little fetuses even get the hiccups!
3. In a poll conducted by Massachusetts Institute of Technology, people were asked to rank the inventions they "couldn't live without." Here is a ranking of their choices: 1. Automobile, 2. Light bulb, 3. Telephone, 4. Television, 5. Aspirin.
4. Did you know that the overseers of the Oxford Dictionary produced a list of 100 key words, one for each year of the century, and "Bobbitt" was their choice for 1993!

## CHAPTER 4

5. The placebo effect can be huge: Anywhere between 35 and 75 percent of patients benefit from taking a sham pill that they think is curative medication (placebo, from Latin, means I shall please). If you stop and think about it, for centuries most medicines and treatments were simply placebos: chanting medicine men in funny masks, barbers' bleedings, charlatans' snake oils, and so on . . . many sufferers got better in spite of them or because of their symbolic value. Even medical researchers are often fooled: If they try a new medication on a patient, and it appears to work, is the success a valid result of the medication, or is it a placebo effect, or did the body's immune system prevail? A difficult distinction to make. Probably as a result of the poor science education offered in the lower grades, about half of Americans have

tried "alternate medical therapies," even though they are aware that there is no empirical proof that the often expensive practices are worthwhile.

On the other hand, capitalizing on the placebo effect can be a valuable, inexpensive tool for the medical profession. A good bedside manner or the laying on of hands can be very effective placebos. Just having a doctor, one in which you have confidence, available if necessary, can be a way to remain untroubled by the little aches and pains that come and go as one ages. Because many active drugs have undesirable side effects, a placebo that works is even better since most often they avoid this down side. Double strength placebos may be the next medical miracle breakthrough.

6. This claim is worthy of a defining footnote. Some of the popular books recently published on the miracles of melatonin suggest that if Ponce de Leon had found the fountain of youth it would have been gushing forth melatonin. Cute, but the truth is that no incontrovertible tests on the purported anti-aging effect on humans have been carried out. The supposition stems mainly from the results of an experiment reported by Dr. Walter Pierpaoli, a researcher working in Italy, who transplanted the pineal glands from 18-month-old mice into 4-month-old mice. On average this shortened the youngsters' life span by one third. When the opposite was done—pineals from 4-month-old mice were transplanted into 18-month-old mice—it extended their lives to an average of 33 months, one-third longer than untreated mice. This study has been touted as proof that melatonin is an anti-aging substance. That is wrong. Whatever caused the change in longevity after pineal transplantation is not known, but it certainly could not have been melatonin because the strain of mice he used does not synthesize melatonin!

CHAPTER 5

7. There are times when being a marine biologist is frustrating (such as having to dive in cold water and sailing in stormy seas) and dangerous (like underwater meetings with large carnivores, or working in pathogenic Avon mud); but being required to collect these shallow-water alga in the Caribbean is one aspect of the job that counters such nasties nicely.

CHAPTER 6

8. William Beebe, a fine naturalist and undersea explorer, once stated, that it is difficult to consider fiddler crabs with the objectivity which all naturalists of worth are supposed to use. I agree with this statement, sometimes they seem somewhat preternatural . . . but I would change his "which" to "that."

CHAPTER 7

9. If you ever want to see one of these tiny flies up close, and insist on having a more realistic view than one gets on the face of a fly swatter, leave a glass of

Beaujolais next to the fruit bowl for an hour. I found out the hard way that the little lushes rush into it like lemmings into the sea.

## CHAPTER 9

10. During a break while working in a salt marsh on a different project, out of curiosity I crossed the two long eyestalks of a captured fiddler crab, one over the other, and tied them in place with thread at the crossover point. The finished design was an X, forcing the right eye to function as a left eye, and vice versa. I set the animal off to one side to let it adjust to the new visual arrangement for a while, and then approached it. Its usual response would have been to run away, but now it ran toward me, I suppose thinking I was on its other side! As soon as I cut away the thread, it again wisely ran away from me.

11. One day while running some learning-ability tests on marine worms, we decided to test fiddler crabs also, to see if they could learn, and if so, how long they would remember the new information. We built a simple T-maze: a straight corridor at the end of which a crab had to turn either right or left. Our desire was to teach them to always turn right by giving a weak electric shock if a left turn was attempted. The first problem was to determine the minimum intensity of a shock needed to be effective. The approach had to be on a trial-and-error basis but eventually we felt we had found the right jolt . . . until we ran into an individual (possibly a different kind of super crab, this one directly akin to Superman who cannot be harmed) who did not respond to the standard shock. We increased the current again, and again, but the crab appeared to ignore it. One more increase finally provoked a response: Sadly, the crab threw off all its legs simultaneously. This was not a "Eureka" moment for us.

    By the end of the day it appeared that some of the test subjects, while never reaching an error-free level of learning, did actually learn to mostly turn right and avoid a potential shock. But when we tested yesterday's B+ students we found not the slightest indication of recall; apparently fiddlers have the memory retention of a slotted spoon.

12. Perhaps you would be interested in hearing about an early attempt by medical experimentalists to get a handle on the soul. Dying patients were carefully weighed just before, and just after, they died. Any difference would be ascribed to the weight of the soul that just departed. Alas, no difference. Conclusion: The soul has no weight!

13. Including a small number of grifters who convinced the world—thankfully for only a few years—to believe in the biorhythm concept, a spurious belief that promised to bring happiness and riches to individuals.

# Suggested Readings Sources

## Chapter 1

Euglena

Palmer, J. 1967. Euglena and the tides. Natural History, 76(2):60–64.

*Commuter Diatom*

Palmer, J. 1975. Biological clocks of the tidal zone. Scientific American, 232:70–79.
Palmer, J. & F. Round. 1967. Persistent, vertical-migration rhythms in benthic microflora. VI. The tidal and diurnal nature of the rhythm in the diatom, *Hantzschia virgata*. Biological Bulletin, 132:44–55.

Convoluta

Palmer, J. 1976. Clock-controlled vertical-migration rhythms in benthic organisms. Pp. 239–255 in: *Biological Rhythms in the Marine Environment*. (De-Coursey, P. ed.). University of South Carolina Press, Columbia, South Carolina.

## Chapter 2

*Cave studies*

Kleitman, N. 1963. *Sleep and Wakefulness*. University of Chicago Press, Chicago.
Siffre, M. 1964. *Beyond Time*. McGraw-Hill, New York.

*Bunker studies of humans*

Wever, R. 1979. *The Circadian System of Man*. Springer- Verlag, New York.

*Modern study of human temperature*

Mackowiak, P. et al. 1992. A reanalysis of human body temperature. Journal of American Medical Association, 268:1578–1580.

*Early birds & night owls*

Chelminski, I. et al. 1999. An analysis of the "eveningness-morningness" dimension in "depressive" college students. Journal of Affective Disorders, 52:19–29.

*Exercise*

Reilly, T. et al. 1997. *Biological Rhythms and Exercise.* Oxford University Press, Oxford, England.

*Blind subject*

Miles, L. et al. 1977. Blind man living in normal society has circadian rhythm of 24.9 hours. Science, 198:421–423.

*Alcohol metabolism*

Wilson, R., et al. 1956. Diurnal variation in rate of alcohol metabolism. Journal of Applied Physiology, 8:556–558.

*Cell division*

Cooper, Z. 1939. A daily rhythm of cell division in neonates. Journal of Investigative Dermatology, 2:289–300.

*Biorhythms*

Palmer, J. 1982. BioRhythm Bunkum. Natural History, 91(10):90–97.

CHAPTER 3

*Cancer*

Hrushesky, W. 1985. Circadian timing of cancer chemotherapy. Science, 228:73–85.

*Asthma*

Smolensky, M. & G. D'Alonzo. 1997. Progress in the chronotherapy of nocturnal asthma. Pp. 205–249 in: *Physiology and Pharmacology of Biological Rhythms.* (Redfern, P & B. Lemmer eds.). Springer-Verlag, Heidelberg.

*General Health*

Smolenski, M & L. Lamberg. 2000. *The Body Clock Guide to Better Health.* Henry Holt & Co., New York.

## CHAPTER 4

*Sham watches*

Lobban, M. 1960. The entrainment of circadian rhythms in man. Cold Spring Harbor Symposium on Quantitative Biology, 25:325–332.

*Phase-shift tables*

Oren, D. et al. 1996. *How to Beat Jet Lag.* H. Holt & Co., New York.

*Illumination of the knee*

Campbell, S. & P. Murphy. 1998. Extraocular circadian phototransduction in humans. Science, 279:396–399.

*Melatonin*

Czeisler, C. 1998. Evidence for melatonin as a circadian phase-shifting agent. Journal of Biological Rhythms, 12:618–623.

Guardiola-Lemaâtre, B. 1998. Toxicology of melatonin. Journal of Biological Rhythms, 12:697–706.

Spitzer, R. et al. 1999. Jet lag: clinical features, validation of new syndrome-specific scale, and lack of response to melatonin in a randomized, double-blind trial. American Journal of Psychiatry, 156:1392–1396.

## CHAPTER 5

Gonyaulax

Hastings, J. et al. 1961. A persistent daily rhythm in photosynthesis. Journal of General Physiology, 45:69–76.

*Venus' Wine Glass*

Sweeney, B. & R. Haxo. 1961. Persistence of a photosynthetic rhythm in enucleated *Acetabularia*. Science, 134:1361–1363.

Mergenhagen, D. & H. Schweiger. 1975. Circadian rhythms in oxygen evolution in cell fragments of *Acetabularia mediterranea*. Experimental Cell Research, 92:127–130.

Parmecium *sex reversal*

Sonneborn, T. & D. Sonneborn. 1958. Some effects of light on the rhythm of mating type changes in stock 232–6 of syngen 2 of *P. multimicronucleatum*. Anatomical Record, 131:601.

Barnett, A. 1965. A circadian rhythm of mating type reversals in *Paramecium multimicornucleatum*. Pp. 305–308 in: *Circadian Clocks* (J. Aschoff, ed.). North-Holland Publications, Amsterdam.

CHAPTER 6

*Fiddler crabs*

Palmer, J. 1990. The rhythmic lives of crabs. BioScience, 40(5):352–358.

*Green crab*

Naylor, E. 1958. Tidal and diurnal rhythms of locomotory activity in *Carcinus maenas*. Journal of Experimental Biology, 35:602–610.
Williams, B. & E. Naylor. 1967. Spontaneously induced rhythm of tidal periodicity in laboratory-reared *Carcinus*. Journal of Experimental Biology, 47:229–234.

*Cockle*

Williams, B. et al. 1993. Comparative studies of tidal rhythms. XIII, Is a clam clock similar to those of other intertidal animals? Marine Biology and Physiology, 13:315–332.

*Geochronometers*

Evans, J. 1975. Growth and micromorphology of two bivalves exhibiting non-daily growth lines. Pp. 119–134 in: *Growth Rhythms and the History of the Earth's Rotation*. (Rosenberg, G.D. & S.K. Runcorn, eds.) John Wiley & Sons, New York.

*Cave cricket*

Simon, R. 1973. Cave cricket activity rhythms and the earth tides. Journal of Interdisciplinary Cycle Research, 4:31–39.

*Palolo worm*

Palmer, J. & J. Goodenough. 1978. Mysterious monthly rhythms. Natural History, 87(10):64–69.
Smetzer, B. 1969. Night of the palolo. Natural History, 87:64–71.

*Human copulation rhythm*

Palmer, J. et al. 1982. Diurnal and weekly, but no lunar rhythms in human copulation. Human Biology, 54:111–121.

*Tidalites*

Sonett, C. et al. 1996. Late proterozoic and paleozoic tides, retreat of the moon, and rotation of the earth. Science, 273:100–104.

CHAPTER 7

*Fruit flies*

Bünning, E. 1976. *The Physiological Clock.* Springer-Verlag, New York.
Pittendrigh, C.S. 1954. On temperature independence in the clock system controlling emergence time in *Drosophila.* Proceedings of the National Academy of Sciences, 40:1018–1029.

*Earthworms*

Bennett, M.F. 1974. *Living clocks in the animal World.* Charles Thomas Publisher, Springfield, Illinois.
Darwin, C. 1881. *The formation of vegetable mold through the action of earthworms, with observations on their habits.* Reprinted by the University of Chicago Press in 1985.

*Bird rhythms*

Binkley, S. *The Clockwork Sparrow.* 1990. Prentice-Hall, Englewood Cliffs, N.J.

*Bird orientation*

Emlen, S. 1975. Migration: orientation and navigation. Pp.129–219 in: *Avian Biology* (Farmer, D. & J. King, eds.). Academic Press, New York.
Goodenough, J., et al. 1993. *Perspectives in Animal Behavior.* See Chapter 11. John Wiley & Sons, New York.

*Penguins*

Emlen, J. & R. Penny. 1966. The navigation of penguins. Scientific American, 215:105–113.

*Bees*

von Frisch, K. 1967. *The Dance Language and Orientation of Bees.* Harvard University Press. Cambridge, Massachusetts.

*Heart-beat rhythms*

Tharp, F. & G. Folk. 1965. Rhythmic changes in rate of the mammalian heart and heat cells during prolonged isolation. Comparative Biochemistry and Physiology, 14:255–273.

## CHAPTER 8

*Sleep-movement rhythms*

Darwin, C. 1881. *The Power and Movement of Plants.* D. Appleton & Co, New York.
Bünning, E. 1957. Endogenous rhythms in plants. Annual Review of Plant Physiology, 7:71–90.
Milnar, A. et al. 1995. Circadian clock mutants in *Arabidopsis* identified by luciferase imaging. Science, 267:1161–1163.

*A survey of plant rhythms*

Sweeney, B. 1987. *Rhythmic Phenomena in Plants.* Academic Press, San Diego.

*Clock control of flowering*

Palmer, J. 1971. The rhythm of the flowers. Natural History, 80(9):64–73.

*Annual rhythm in germination*

Bünning, E. 1949. *Zur Physiologie der endognen Jamresrhythmic in Pflanzen, speziell in Samen.* Z. Naturforsch (B), 4:167–176.

## CHAPTER 9

*The suprachiasmatic nuclei (SCN)*

DeCoursey, P. et al. 1997. Circadian performance of suprachiasmatic nuclei (SCN)-lesioned antelope ground squirrels in a desert enclosure. Physiology and Behavior,62:1099–1108.
Klein, D. et al. (editor). 1991. *Suprachiasmatic Nucleus: The Mind's Clock.* Oxford University Press, New York.
Takahashi, J & M. Menaker. 1982. Role of the suprachiasmatic nuclei in the circadian system of the house sparrow, *Passer domesticus.* Journal of Neurosciences, 2:815–828.
Takahashi, J. 1995. Molecular neurobiology and genetics of circadian rhythms in mammals. Annual Review of Neuroscience, 18:531–553.

*Tidal rhythm clocks*

Palmer, J. 1995. *The Biological Rhythms and Clocks of Intertidal Animals.* Oxford University Press, New York.
Palmer, J. 2000. The clocks controlling the tide-associated rhythms of intertidal animals. BioEssays, 22:32–37.

*The pineal*

Gaston, S. & M. Menaker. 1968. Pineal function: the biological clock in the sparrow? Science, 160:1125–127.

*The molecular clock*

Dunlap, J. 1999. Molecular bases for circadian clocks. Cell, 96:271–290.
Green, C. 1998. How cells tell time. Trends in Cell Biology, 8:224–230.
Tei, Hajime et al. 1997. A sequence comparison of the amino acid products of three per genes. Nature, 389:512–516.

# FIGURE & DRAWING CREDITS

*Euglena* on page 3. Palmer, J.D. 1967. Euglena and the tides. *Natural History,* 76(2):60–64.

Commuter diatom on page 5. Palmer, J.D. 1974. An Introduction to Biology Rhythms. Academic Press.

Figure 1.1. Palmer, J.D. 1975. Biological clocks in the tide zone. Scientific American, 232:70–79.

*Convoluta* on page 8. Keeble, F. 1910. *Plant-animals: A Study in Symbosiosis.* Cambridge University Press.

Figure 2.1. Modified from: Siffre, M. 1964. *Beyond Time.* McGraw-Hill. New York.

Figure 2.2. Wever, R.A. 1976. *The Circadian Systems in Man: Results of Experiments Under Temporal Isolation.* Springer-Verlag New York.

Figure 2.3. Modified from: Miles, L. et al. 1977. Blind man living in normal society has circadian rhythm of 24.9 hours. Science, 198:421–423.

Figure 2.4. Colquhoun, W.P. ed. 1971. *Biological Rhythms and Human Performance.* Academic Press. New York.

Figure 2.5. Wever, R.A. 1976. *The Circadian Systems in Man: Results of Experiments Under Temporal Isolation.* Springer-Verlag New York.

Figure 2.6. Wever, R.A. 1976. *The Circadian Systems in Man: Results of Experiments Under Temporal Isolation.* Springer-Verlag New York.

Figure 2.7. Wilson, H.L. et al. 1956. Diurnal variation in rate of alcohol metabolism. Journal of Applied Physiology, 8:556–558.

Figure 4.1. Palmer, J.D. 1983. *Human Biological Rhythms.* Carolina Scientific Publications, North Carolina.

Figure 5.1 Modified from: Sweeney, B. & R. Haxo. 1961. Persistence of a photosynthetic rhythm in enucleated *Acetabularia.* Science, 134:1361–1363.

Fidler crab on page 77. Margaret Nutting, Amherst, MA

Figure 6.1. Brown, F.A. et al. 1970. *The Biological Clock: Two views.* Academic Press, New York.

Figure 6.2. Palmer, J.D. 1974. *Biologial Clocks in Marine Organisms.* Wiley-Interscience. New York.

Figure 6.3. Brown, F.A. et al. 1970. *The Biological Clock: Two views.* Academic Press, New York.

Samoan Palolo worm on page 87. Palmer, J.D. 1995. The Biological Rhythms and Clocks of Intertidal Animals. Oxford University Press. New York.

Eating Samoan Palolo Worm on page 88. Smitzer, B. 1969. Night of the palolo. Natural History, 87:64–71.

Figure 6.4. Palmer, J.D. et al. 1982. Diurnal and weekly, but no lunar rhythm in human copulation. Human Biology, 54:111–121.

Figure 7.1. Goodenough, J.E. et al. 1977. The biological clock: regulating the pulse of life. *Sci. Teacher,* 44:31–34.

Figure 7.2. Palmer, J.D. *An Introduction to Biological Rhythms.* Academic Press. New York.

Figure 7.3. ibid

Figure 7.4 Bennett, M.F. 1974. *Living clocks in the animal world.* Charles Thomas Publ.

Figure 7.5. Dowse, H.B. 1971. Ph.D Thesis, Univ. Massachusetts

Figure 8.1. Palmer, J.D. 1984. *Biological Rhythms and Living clocks.* Carolina Biology Publ.

Figure 8.2. Palmer, J.D. 1976. *An Introduction to Biological Rhythms.* Academic Press. New York.

Figure 8.3. ibid

Figure 8.4. ibid

Figure 9.1. Truman, J. 1971. The role of the brain in the ecdysis rhythm of silkmoths: comparison with the photoperiodic termination of diapause. Pp. 483–504. In: *Biochronometry.* National Academy of Sciences. Washington, D.C.

Figure 9.2. Palmer, J.D. 1974. *Biological Clocks in Marine Organisms.* Wiley-Interscience Publ. New York.

Figure 9.3. Palmer, J.D. 1995. The Biological Rhythm and Clocks of Intertidal Animals. Oxford University Press. New York

Amino acid sequences on page 142. Courtesy of Dr. Hajime Tei.

# INDEX